STUDIES IN ECONOMIC AND SOCIAL

This series, specially commissioned by the Economic History Society, provides a guide to the current interpretations of the key themes of economic and social history in which advances have recently been made or in which there has been significant debate.

Originally entitled 'Studies in Economic History', in 1974 the series had its scope extended to include topics in social history, and the new series title, 'Studies in Economic and Social History', signalises this development.

The series gives readers access to the best work done, helps them to draw their own conclusions in major fields of study, and by means of the critical bibliography in each book guides them in the selection of further reading. The aim is to provide a springboard to further work rather than a set of pre-packaged conclusions or short-cuts.

ECONOMIC HISTORY SOCIETY

The Economic History Society, which numbers over 3000 members, publishes the *Economic History Review* four times a year (free to members) and holds an annual conference. Enquiries about membership should be addressed to the Assistant Secretary, Economic History Society, Peterhouse, Cambridge. Full-time students may join at special rates.

STUDIES IN ECONOMIC AND SOCIAL HISTORY

Edited for the Economic History Society by L. A. Clarkson

PUBLISHED

OTHER TITLES ARE IN PREPARATION

The British Iron Industry 1700–1850

Prepared for
The Economic History Society by

J. R. HARRIS

Professor of Economic History
University of Birmingham

M
MACMILLAN
EDUCATION

First published 1988

Published by
MACMILLAN EDUCATION LTD
Houndmills, Basingstoke, Hampshire RG21 2XS
and London
Companies and representatives
throughout the world

Printed in China

British Library Cataloguing in Publication Data
Harris, J. R.
The British iron industry, 1700–1850.—
(Studies in economic and social history).
1. Iron industry and trade—Great Britain
—History   2. Steel industry and trade—
Great Britain—History
I. Title II. Economic History Society
III. Series
338.4'76691'0941   HD9521.5
ISBN 0–333–33979–7

Contents

Acknowledgements

I have had a great deal of generous help and advice from friends who have read partial or complete drafts of this work; I hope I have taken full advantage of their assistance and for any remaining faults I am all the more to blame. I have in particular to thank Kenneth Barraclough, David Crossley, Keith Gale, Charles Hyde, Barrie Trinder and Ronnie Tylecote. Between them they have saved me from many errors of fact or emphasis. As a colleague Barrie has been particularly accessible and has been able to advise on the drawings and the note on industrial archaeology. Another colleague, Eric Hopkins, and Paul Harris have read the script with the needs of the student reader in mind; having had a long connection with the promoting of this series it has always seemed to me that they are the important customers. Sue Kennedy has transformed a bad manuscript into a neat and accurate typescript. The drawings have been prepared at Ironbridge by Sue Isaacs; participating in the work of the Ironbridge Institute has been the happiest background to preparing this little work.

J.R.H.

Editor's Preface

When this series was established in 1968 the first editor, the late Professor M. W. Flinn, laid down three guiding principles. The books should be concerned with important fields of economic history; they should be surveys of the current state of scholarship rather than a vehicle for the specialist views of the authors, and above all, they were to be introductions to their subject and not 'a set of pre-packaged conclusions'. These aims were admirably fulfilled by Professor Flinn and by his successor, Professor T. C. Smout, who took over the series in 1977. As it passes to its third editor and approaches its third decade, the principles remain the same.

Nevertheless, times change, even though principles do not. The series was launched when the study of economic history was burgeoning and new findings and fresh interpretations were threatening to overwhelm students – and sometimes their teachers. The series has expanded its scope, particularly in the area of social history – although the distinction between 'economic' and 'social' is sometimes hard to recognise and even more difficult to sustain. It has also extended geographically; its roots remain firmly British, but an increasing number of titles is concerned with the economic and social history of the wider world. However, some of the early titles can no longer claim to be introductions to the current state of scholarship; and the discipline as a whole lacks the heady growth of the 1960s and early 1970s. To overcome the first problem a number of new editions, or entirely new works, have been commissioned – some have already appeared. To deal with the second, the aim remains to publish up-to-date introductions to important areas of debate. If the series can demonstrate to students and their teachers the importance of the discipline of economic and social history and excite its further study, it will continue the task so ably begun by its first two editors.

The Queen's University of Belfast   L. A. CLARKSON
General Editor

Note on References

References in the text within square brackets relate to the number-ed items listed in the Bibliography, followed, where necessary, by the page numbers in italics, for example [1, 7–9].

1 Introduction

(i) THE BLOOMERY

Iron was first made in the Near East, possibly as early as 1500 BC, and on a regular basis from 1000 BC, and it reached Britain with Celtic immigrants around 450 to 500 BC [83, *40*; 72, *4*]. There are three main varieties of iron whose chief chemical distinction is their carbon content. Wrought iron is tough and most malleable when hot and capable of being forged by the smith into a range of goods. Down to about 1500 when the blast furnace arrived in Britain, wrought iron was virtually the only kind produced here. Some wrought iron was subsequently given a steeled surface by heating in contact with charcoal or other carbonaceous material, a very little may have had enough carbon penetration by more protracted heating to produce some steel [72, *116–21*]. But steel, sometimes produced by more developed processes, was often imported from those parts of Europe (particularly Westphalia) which produced purer iron than Britain did; in any case its high price severely restricted its uses. Cast iron as yet had no commercial value and was produced only inadvertently. Wrought iron had a very low carbon content, cast iron a much higher one, steel being intermediate.

In principle the production of iron in the late Middle Ages was little different from that of earlier periods. Wrought iron was produced in a *bloomery*. This may have been a roughly circular hollow, sometimes clay lined, which could vary in width and depth from between perhaps 18 inches to 3 feet deep and 1 to 4 feet wide or, occasionally, a furnace with a short shaft. Iron ore, bedded in and covered by charcoal, was placed in the interior and a hot fire maintained by one or more pairs of bellows, usually foot-operated, the air entering the furnace through a duct. The ore never liquefied but acquired the character of a stiff pasty lump which, when the master operator (the *bloomer*) judged the appropriate time had arrived, was taken out of the hearth and hammered to express slag

11

and to produce a consistency of wrought iron suitable for the smith. Reheating was needed during the hammering operation. Wrought iron was thus produced *directly* from the ore. The archaeological evidence of medieval bloomeries, however, is limited and not easy to interpret. A type with a clay dome, known in earlier periods, has been identified as existing then. Some bloomeries were so designed as to allow slag to be run off. Some bloomeries were permanent with lasting buildings, others were operated for shorter periods under temporary cover by itinerant workers. From about the twelfth century it became common to do the reduction in a first or *bloom hearth* and the reheating in a second or *string hearth*. Possibly soon after 1200, but commonly by around 1400, a waterwheel was increasingly resorted to to operate the bellows for one or both hearths [72, *Ch. VIII*; 83, *65*; 84, *passim*]. In the late Middle Ages there was an increase in the dimensions and the production of the bloomery (the water-powered is called a *bloomsmithy* by some authors – [6, *25*]). The bloom seems to have rapidly increased in size from around 30 lb in the fourteenth century to around 100 or even 200 lb in some cases in the fifteenth. Annual output of a bloomery rose from perhaps a couple of tons a year to as much as 25 tons [72, *133–40*], though up to a fourth in weight could be lost in the reheating and hammering process. The acceleration in the increase in bloom size and furnace output in the fifteenth century is almost certainly related to the use of water power for a second purpose, the hammering of the bloom. This would have briefly anticipated the major change of the introduction of the indirect method before 1500. Subsequently water power continued to be applied to hammers at large bloomeries as well as to the new blast furnaces and their forges.

The bloomery thus did not disappear with the introduction of the blast furnace and its allied but separate forges, however significant the new development was. Even if it was wasteful in iron lost in slags, the bloomery produced in one stage a good wrought iron, and its capital requirements were modest, particularly for the smaller furnaces. Where population was low and local demand scattered, production in small quantities or intermittently fitted the economic requirements. For instance Cumbria saw the development of something like 20 bloomeries in the seventeenth century before the first blast furnace there arrived in the 1690s [64, *28*]. Even a few miles from the rising port of Liverpool a bloomery seems to have been worked by itinerant workers as late as 1712 [10], and another is noted in Lancashire in 1756. Many bloomeries were worked in Scotland between the sixteenth and eighteenth centuries [72, *151*– *2*; 84, *187*]. One would naturally expect the medieval bloomeries to have been scattered and to have appeared in many places where modest amounts of ironstone and charcoal could be found, but factors such as high-quality ore quickly made the Forest of Dean and the Wye Valley the principal region, whence there was some extension in S. Wales; another significant area was between Sheffield and Rotherham and northwards towards Leeds. There were other concentrations of medieval bloomeries in the following areas; in the Weald; in Cumberland and the Furness peninsula; to the west of Scarborough and in the Cleveland Hills; in North Derbyshire and between Sheffield and Chesterfield, in Cheshire; in parts of what later became the 'Black Country'. In Scotland the most important site was near Elgin. There were at least 150 bloomery furnaces in Britain in the last half of the thirteenth century, probably 350 sites occupied at some time in the Middle Ages. The increase in ouput with the coming of water-powered bellows and hammer helps explain the late continuance of bloomeries which could eventually produce as much as 45 tons a year [72, *98 seq.*, *139*– *40*, *149*]. The bloomsmithy with water-powered blowing and hammering did not remain technically static, but from its bowl furnace developed until it was remarkably similar to the forge stage of the indirect process to which we now turn.

(ii) THE ARRIVAL AND NATURE OF THE INDIRECT PROCESS

The first part of the period 1680 to 1830, that down to 1750, was dominated by the so-called indirect process introduced into England just before 1500, which gradually eliminated the bloomery and in any case fairly rapidly superseded it as the major producing process. It is called an *indirect* process because while the main product continued to be *wrought* iron, this was now arrived at by first producing *pig* iron, in fact a cast iron, from ore which was raised to a high temperature (*c*. 1400°C) in a *blast furnace*, which completely liquefied it. The iron having been run off and solidified was then taken to a *forge* where it was *refined* on a charcoal hearth under an air draught from bellows. It was hammered to consolidate it, reheated in a *chafery* hearth and hammered into wrought iron *bar*. When

13

writers refer to the *charcoal iron industry* they refer to this indirect process at the period when charcoal was the fuel, before coal was employed in the smelting of iron [72, *Chs X, XII*; 27, *20–6*; 83, *82–4*]. This is a little confusing to the student, because of course the bloomery was also a method of making iron with charcoal.

The most obvious feature of the new system was the *blast furnace*. Versions had appeared in China by the fifth century BC (and there coal was sometimes used as fuel from the eleventh century AD [60; 61, *101 seq*; 83, *48, 69, 85*], but it is not known if any knowledge of it percolated to the west. It seems more likely that there was a spontaneous western invention. There is recent archaeological evidence of a blast furnace in Sweden not later than the mid fourteenth century [49, *51 seq*; 30, *110–15*], but it seems certain that the point of derivation of the blast furnace as it came to England was the Liège area of modern Belgium, whence it spread into northern France and thence across the Channel to the Weald [8, *99* and map, *104*; 72, *163–5*]. We do not yet know of any connection between the Swedish development and the appearance of the blast furnace in Belgium.

The blast furnace as it arrived in England was a tower around 15 feet high which eventually achieved a classic form of about 25 feet in the charcoal-iron era. The furnace was periodically fed from the top with charcoal and iron ore (with some limestone added as a flux if necessary). Soon it became normal to build it close to a bank of ground from which the top could be reached by a ramp or *bridge* for charging purposes. The full technology is complex and a matter for specialists, but briefly in Britain the interior of the furnace came to have a profile like two pyramidal cones connected at their base, the slopes of the lower inverted one forming the supporting surfaces (*boshes*) which bore the weight of the charge. At the bottom was a *hearth* or *crucible* of smaller dimensions in which the liquefied iron settled, so aranged that from an upper orifice slag floating on the iron could be tapped, and from a lower one liquefied iron could be periodically run off into a moulded sand bed. Here it solidified into a long length called the *sow* and (in its classic form) lesser lengths at right angles to the main sow, appropriately called *pigs*. While the bloomery had intermittently produced individual blooms, and the production process had been *discontinuous* as the blooms were taken from the hearth, hammered and reheated, the blast furnace was a *continuous* process in which an extended *campaign* of months and eventually sometimes more than a year could be maintained. If supply of materials and demand for the product continued good, only a lack of water power or heat damage to the interior would enforce a break in production. Water power of course was required for bellows – in the mature form of the charcoal blast furnace double bellows – which could keep up a sufficiently powerful and continuous blast as they were actuated by cams on a shaft turned by the water wheel. To ensure operation with adequate water power the furnace was so sited that a quantity of water could be pent up behind a dam to form a *furnace pond* [72, *Ch. XIV*; 27 *20 seq*; 43, *8 seq*].

While it was the first stage of a two-stage process, the blast furnace meant at once an increased scale of production, for a furnace could even early in the sixteenth century yield 200 tons a year [18, *273*]. Of course the consumption of charcoal was much larger in the blast furnace than in the bloomery, and probably as an extension of a practice previously developing with larger bloomsmithies, arrangements needed to be made to grow wood as a crop in the neighbourhood. This meant that it was difficult to cluster blast furnaces together. For the same reason it could also be difficult to put the forges which converted the pig iron to wrought iron in the second stage of the new indirect process close to the blast furnace. But there was another constraint, the second-stage hearths needed water-power, not only for bellows as did the blast furnace, but also to work hammers, so unless the head of water at the furnace was superabundant, the second stage tended for that reason also to be at some distance. Occasionally former bloomeries may have been taken over and adapted for the purpose. This second stage at which the pig (cast) iron was converted to wrought iron was called the *forge* stage. The forge consisted firstly of the *finery*, a hearth not unlike a blacksmith's where the pig iron was heated and stirred, while a blast of air from water-forced bellows was directed on it. The oxygen in the blast combined with the carbon in the iron and, as the carbon was reduced, the cast iron was turned into a pasty wrought iron which was then taken to a water-powered hammer. The hammer shaft was balanced on a fulcrum so that its tail end could be depressed and released by cams operated from the shaft of a water wheel. The effectiveness of the hammer was increased by contact with a *spring-beam* above it. The finery and the hammering resulted in a block of perhaps half-a-hundredweight of wrought iron. This was reheated in a separate *chafery* hearth and again taken to another hammer; this chafery operation produced no chemical change in the iron but worked the iron into the form in which it was normally sold, that is *bar* or wrought iron. The forge stage had many variations between the European producers. As would be expected from the origins of the indirect process in Britain, we used the two-hearth Walloon process rather than the German single-hearth type and this is reflected in our technical terms, which are often of French origin. (Some idea of the complexity can be gained from [21, *63 seq*]; see also [72, *Ch. XVI*; 27, *21–5*].)

The first record of a blast furnace in England is at Newbridge in the forest of Ashdown in the Weald of Sussex, about the middle of the 1490s. Newbridge was a royal manor, and the furnace was early called upon to produce not only pig iron for conversion to wrought iron, but also iron castings, principally in the form of shot for artillery. Perhaps this preoccupation caused the spread of the blast furnace to be fairly slow, there being nine in Sussex by 1542, when immigrant French workers were still prominent. Such French workers were probably critical in the first major casting of iron cannon near Buxted in 1543 [72, *165 seq, 170–2*; 17, *123–5*]. The dependence of England on transferred Continental technology is clear at this period, but in the casting of cannon the English soon came to surpass their European teachers and rivals. In other respects the English technical vocabulary of the indirect process with words like 'tuyere', 'chafery' sufficiently shows the French origin.

For the introduction of one further piece of technology Britain was again dependent on the Continent. The *rolling and slitting mill* was apparently introduced from Flanders in the 1580s and was certainly operative about 1590. In the developed form which it reached in the next century pairs of rotating water-powered iron shafts had portions formed as rolls to flatten out hot iron bar passed through them, and then the flattened iron was passed between sets of rotating discs on the same shafts which cut them into thin rods, highly suitable for many of those working with iron, notably nailmakers. The rolling and slitting mill may be said to have completed the technology of the charcoal iron period, though like the blast furnace it continued on into the age of coal and steam power [72, *304 seq*; 27, *28*].

The new indirect system first extended within the Sussex Weald, searching out the better ores and locations near the coast with sea

CHARCOAL BLAST FURNACE Cut Open To Show Interior

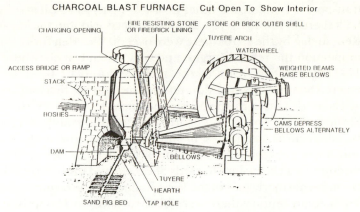

FIRE RESISTING STONE OR FIREBRICK LINING
STONE OR BRICK OUTER SHELL
CHARGING OPENING
TUYERE ARCH
WATERWHEEL
WEIGHTED BEAMS RAISE BELLOWS
ACCESS BRIDGE OR RAMP
STACK
BOSHES
CAMS DEPRESS BELLOWS ALTERNATELY
DAM
BELLOWS
TUYERE
HEARTH
SAND PIG BED
TAP HOLE

(From a diagram copyright: N. Cossons)

PUDDLING FURNACE
Sectional Elevation

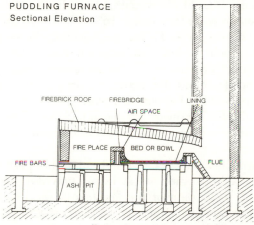

FIREBRICK ROOF   FIREBRIDGE   LINING
AIR SPACE
FIRE PLACE   BED OR BOWL
FIRE BARS   FLUE
ASH PIT

(Redrawn from a diagram copyright: W.K.V.Gale)

ROLLING MILL

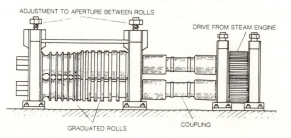

ADJUSTMENT TO APERTURE BETWEEN ROLLS
DRIVE FROM STEAM ENGINE
GRADUATED ROLLS   COUPLING

17

transport to London and other ports. As many as 15 furnaces were built there in the 1540s [18, *277*]. It then spread into the Weald of Kent so that by the 1570s there were over 50 Wealden furnaces, not to mention the related forges, and only a handful elsewhere [72, *174*]. By the first years of the next decade blast furnaces had entered South Wales and Monmouthshire, and the first had arrived at Cannock Chase in the West Midlands, followed by others in Shropshire, in South and North Staffordshire and as close to Birmingham as Aston, West Bromwich and Halesowen. Derbyshire and North Yorkshire came into the picture with a band of furnaces through East Derbyshire towards Sheffield and there were a few in the more northerly part of the West Riding. By the beginning of the seventeenth century there were blast furnaces in Hampshire [72, *179–83*]. The first furnaces on the borders of the Forest of Dean appeared as early as the 1570s [37, *8*] but the really rapid extension there, replacing the old bloomeries, came shortly after 1600 and there were said to be 11 furnaces in 1633 [37, *13*]. At the end of the seventeenth century the Forest produced a large proportion (perhaps as much as half) of English iron. The high-grade haematite ore of Lancashire and Cumberland was exported to furnaces in Scotland and the north of Northern Ireland, and the area's only iron works of any permanence were bloomeries before the celebrated Backbarrow works began in 1711. This was notable as the last furnace to operate with charcoal, continuing to use it until 1920.

While John Fuller's celebrated list of 1717 is not perfect, it gives a sufficiently good impression of the distribution of the charcoal iron industry at the beginning of the eighteenth century. It shows a heavy concentration of the industry in the Forest of Dean and the West Midlands with outliers in South Wales, North Wales and Cheshire, isolated operations in North Lancashire, a band of works between Derby and Leeds and the still active, but proportionally declined, Wealden area, with its smaller works [72, *193*].

2 Historical Debate on the Charcoal Iron Era

(i) GENERAL

We now enter the period where, if controversy be the criterion, the interest of the iron industry must be at its greatest. Controversy centres on a number of its aspects: on output, whether it declined, remained stable or rose during the late seventeenth and early eighteenth centuries; whether the industry was at the limits of available fuel and water power and constrained to relocate by their costs; whether labour costs were the most adverse factor; what were the reasons for foreign penetration of the British market and what the foreign share was. It will, however, be easier to understand the much debated forces acting on the industry of the late seventeenth and early eighteenth centuries if we first briefly look at how the industry operated.

The main product of the industry was wrought iron of various grades, from the high quality needed for the wire-drawer to that for the common work of the blacksmith or that which would satisfy the nailmaker; the last would be too *coldshort* (brittle when cold) for many other users. Cast iron from the first stage of the indirect process was only used for a few products like firebacks. It is commonly said to account for only 5 per cent of production [43, *18*], though this must have varied between peacetime and wartime, for in war there was a demand for cast iron guns and shot. While landowners including the Crown were important in the early stages of the industry's establishment, often because they had woods whose value would be raised by the demand for charcoal, there was a progressive move to the leasing and operation of iron works by more specialised people [19, *68*], active entrepreneurs with a greater technical and market knowledge, who came to be called *ironmasters*. Furnaces and forges were sited with regard to ore supplies, to water power and to the woods supplying charcoal. Ironmasters would construct them, not with the short-term policy of using up local wood supplies and then removing their works, but fixing with local landowners arrangements for a permanent supply [23, *148 seq*]. This could be ensured by the planned replanting of woods in a cycle of growth to produce coppices of around 20 years maturity, which supplied the best wood for charcoal. Consequently the size of individual furnaces and forges and their output tended to remain fairly static, for they had been set up with the wood resources of the area in mind, often in a district where there might be some alternative and competitive uses for land, particularly agricultural. The ironmaster could not extend his charcoal demands too far afield. This was not only a matter of transport costs, largely made up of the wages of labour and the keep of animals, but charcoal moved over any distance degenerated because it was so fragile and friable, crumbling from pieces of a size fit for the furnace to a useless dust. Ironmasters were prepared to carry it five miles, but ten or twelve miles was highly undesirable and only rarely exceeded [23, *150*; 32, *606*].

Iron ore mining, normally very shallow, did not often present critical problems, but the quality of ore did, and so there were variations in the resulting pig iron because of the deleterious presence of (for example) phosphorus or sulphur. This was a powerful reason for ironmasters operating a variety of furnaces and forges, for different qualities of pig iron could be mixed at the forges to produce the qualities of wrought iron in current demand. Forges nearest to a market (for instance Birmingham) could naturally respond most quickly to changing demands. Clearly furnaces, forges and markets which were connected by cheap water transport had advantages, hence the great importance of the River Severn linking the quality 'tough' iron of Dean through the river port of Bewdley with the forges serving the Birmingham area. In such forges on the Stour and other local rivers it was mixed with the 'coldshort' iron of Staffordshire and Shropshire [46, *333–4*; 74, *30–1*]. The Birmingham region was a great iron consumer because of the cheap coal which was used in reheating wrought iron to work it into goods, including nails. Over time it was found possible to employ coal further back in the chain of production, at the rolling and slitting mills, and even in the chafery of the forge stage [43, *80– 2*; 47, *71*], and this gave ironmasters a noticeable saving.

The vertical connection traceable from ore mining to slitting mill, and the bringing together of the product of scattered works to supply major markets, naturally led to groupings of ironmasters to operate partnerships, each generally based on leading families, which could control a complex trade. Several are famous, including the Knight partnership on the Stour and the Spencers in South Yorkshire [22, *90 seq*; 67; *165 seq*], but the business aspect is perhaps best exemplifed by the Foleys at the height of their influence. The third generation of the family was dominated by Thomas II, Philip and Paul Foley. Their grandfather Richard had founded iron-works, extended by their father Thomas I, whose trade was facilitated by their uncle Robert, a great *ironmonger* (i.e. iron merchant), of Stourbridge, a key centre of the iron trade. Thomas Foley I had inherited a group of works from his father, along the Stour. He added works in Gloucester, Hereford, Worcester, Staffordshire, Shropshire and Monmouth, leaving at his death in 1677 furnaces, forges, slitting mills, wire works and stores, 'an integrated network stretching from the lower Wye to . . . northern Staffordshire'. Well before his death he had begun to divide his works between his sons, creating 'autonomous segments', roughly for Monmouthshire, for Dean, and for the Stour Valley, the divisions probably being due to the difficulty of managing the full, dispersed enterprise.

Eventually, after years of decentralisation, the three sons recreated an 'ironworks in partnership' in 1692. They had followed their father and grandfather by making partnerships with others to run particular works. Now they formed a 'supervisory council of partners' to manage the whole, appointing John Wheeler to be effective managing director, and devolving the shareholding so that together the brothers owned only a third. The active brothers, Paul and Philip, respectively controlled mainly furnaces in the case of the former and forges in the case of the latter, so that despite some enmity they eventually had to cooperate, even though for a period before 1692 they had each tried to expand their own works to seek independence of the other. It was the marketing problems that this independence had caused which had led to the peacemaking of 1692. The partnership then agreed was to be Paul and Philip Foley, John and Richard Wheeler and Richard Avenant, together in-terested in 'melting, making, slitting and disposing of pig, barr and rod iron' at four furnaces, three forges and a slitting mill, with provision for taking in works leased out to other associates as the leases fell in. They established the value of the first group of works at £39,000 and each agreed not to enter any other iron partnership without the consent of all. Even so, the interests of the partners and the demands of trade led to a series of further reorganisations. The partnership gradually withdrew from the Stour forges, subsequently the base of another powerful group in the industry headed by the Knight family, but the 1692 reorganisation nevertheless was the main foundation for the continuation of the Foley interest for three-quarters of a century, after which they departed the industry with the demise of charcoal iron.

At the turn of the eighteenth century, apart from the 'iron works in partnership' the Foleys had separate partnerships in several furnaces in North Staffordshire and Cheshire, a Flintshire forge and a Sussex furnace and forge, while they had an unknown interest in a group of furnaces and forges in Derbyshire and Nottinghamshire. Despite their dispersed nature the works were largely worked with an integrated policy. For instance, Dean pig iron gave tough merchant iron as its product at their Stour forges, while a lower grade (best mill bar) consisted of a mixture of Dean and coldshort iron, while ordinary mill bar, largely coldshort, could also be provided. The purpose of it all was largely that of 'supplying the midland market and in particular . . . Birmingham and the industrial towns and villages of the adjacent coalfield' [47, *322 seq*; 74, *19 seq*]. The above has given a rough and highly condensed account of one of the most important of the charcoal iron groups operated by integrated partnerships. Space forbids us to examine others with major similarities but much difference in detail, like those of the Knights and the Spencers.

(ii) THE QUESTION OF OUTPUT

At this point we must examine the important matter of the output of the iron industry in the late seventeenth and early eighteenth centuries. In a few prefatory pages to his celebrated study of the eighteenth-century iron industry, first published in 1924, Ashton gave some powerful impressions of the later seventeenth- and early eighteenth-century situation which would now be regarded as misleading. They were all the more forceful because of the striking phrases which he was so good at penning. He took some limited evidence of government restraint on the iron industry, mainly in the late sixteenth century, as indication of a serious shortage of fuel, and he interpreted the spread of ironworks away from the Weald into the South West, Midlands and North as impelled by a desperate search for fuel. 'As the hunger for fuel increased iron masters were forced to migrate into more remote lands; salvation could only be found in solitude, and the industry of smelting and refining was literally fleeing to the wilderness to escape destruction' [3, *15*]. He acknowledged that in some areas afforestation and careful coppicing were practised, but regarded this as palliative and wood resources as able at best to increase in arithmetical, as ironmasters' demands increased in geometrical, proportion. The double need for ironmasters to find water power as well as fuel made, he thought, a crippling constraint, so that the industry cringed before a 'tyranny of wood and water' [3, *22*]. On very limited quantitative evidence he concluded that the iron industry declined significantly from the middle of the seventeenth century and that the figures of around 20,000 tons of total national output about 1720 represented a serious fall from a seventeenth-century peak [3, *13* and *Ch. I passim*; 23, *144*]. Other writers subsequently adopted this view.

The first major revision of this interpretation came at the end of the 1950s. M. W. Flinn examined the growth of the English iron industry between 1660 and 1760, the latter date roughly indicating the timing of the great expansion of the iron industry based on coal fuel. Flinn claimed that Ashton and others concentrated excessively on the Weald area and while this area did decline, others indubitably rose. He pointed out that some of the data used by earlier writers had come from parliamentary material and contemporary pamphlet evidence in the first half of the eighteenth century when there had been controversy about the industry, sometimes during periods of restrictions on imports or of other constraints on supply. It had then been in the interest of particular groups within the industry to suggest low home production figures, and of course there were then no reliable official statistics against which to check them. An attempt by E. W. Hulme in the 1920s to produce a statistical picture of the trade at this period had omitted a significant number of furnaces and forges, and the estimates he made of output of works were sometimes founded on untypical years. Even for the Weald there had been a late seventeenth-century revival. The main fact was that many new blast furnaces were built over 1660–1760 and a few earlier ones rebuilt, and Flinn argued that investment in 43 furnaces hardly squared with a declining industry, while 29 new forges were also known, certainly an underestimate, as the output of the new furnaces was mostly sent to forges. Flinn believed that new capacity had added between 15,000 and 20,000 tons to annual pig output. Of course the production of some abandoned furnaces had to be subtracted, but this still allowed for an increase of 10,000 tons of pig over the century. The furnaces going out of production included a high proportion of small ones – some in the Weald had only made 150 tons a year – while newer ones tended to be larger, capable of making 700–1000 tons in a year. However, the loss of metal in the conversion of pig iron to wrought iron would reduce the additional 10,000 tons to little over 7000, if we ignore the very small cast iron market which used uconverted pig [23, *144 seq*].

Those writers who had suggested a declining industry had argued that there had been a shortage of charcoal, and Flinn attacked this idea too. He stressed renewal of timber resources by coppicing in most iron-producing districts and the fact that this generally gave cordwood of good quality at a cost acceptable to the ironmasters. As the industry grew slowly, new furnaces were admittedly built in new areas. But this was because individual furnaces, due to transport cost and the fragility of charcoal, could only draw supplies from a limited area. If existing ironmasters, in a district where accessible wood supplies were already well employed, all decided to increase output substantially by rebuilding larger furnaces, or if incomers built new furnaces in such areas, they would of course force up fuel prices. So ironmasters would normally avoid such practices. It was more sensible to set up new furnaces in a new area possessing ore and water power but where woods were untouched, and where landowners would welcome revenue from planting and selling coppice wood. Flinn thought that lack of enough water was a much more common handicap to ironworks in this period than was shortage and dearness of charcoal [23, *148 seq*].

The thesis of charcoal shortage also received a strong attack from Hammersley, who believed the relatively less wooded state of Britain in the nineteenth century, compared with many continental countries, had led to a false inference about fuel problems for the iron industry in the two previous centuries. He showed conclusively that the idea of decline had been partly founded on exaggerated or misinterpreted figures, which had given exaggerated totals of

40,000 tons and upwards of bar iron production for the early seventeenth century. He showed that ironmasters were not continually forced to close furnaces and re-erect elsewhere; three-tenths of all those furnaces between 1520 and 1749 of which record survives remained active for over a hundred years, well over half for more than fifty. The spread of the industry out of the Weald was due to the attractions of a profitable industry to landowners elsewhere rather than to a desperate search for scarce fuel. Like Flinn he suggests an increase of about 10,000 tons in pig production between the 1680s and the middle of the eighteenth century [32, 593 seq]. Hammersley makes a bold attempt to estimate the extent of timber supply. The increased efficiency of the operation of furnaces and forges between 1540 and 1750 economised on charcoal, fuel consumption in smelting fell by half, that at the forge by between two-fifths and a half, while less pig iron was lost in the conversion to wrought iron at the forge stage. Putting maximum production under the indirect system at 35,000 tons of pig or 27,000 tons of wrought iron, Hammersley believed that the maximum consumption of wood under that system of production was less than 60 million cubic feet a year. Consequently 650,000 acres of woodland or less than 2 per cent of the land surface of England and Wales could have maintained the industry in perpetuity. However he candidly admits that wood as a crop could only profitably use what to the landowner was marginal land, as it yielded an income 'far below the return on pasture or other crops' [32, 603 seq, 609]. Again, he probably does not emphasise enough the significance of rises in the price of charcoal in the seventeenth century, particularly between 1630 and 1670, as an adverse influence on the industry [79].

The absolute amount of acreage either already in woodland, or capable of being planted in coppices, and for which there was not a more profitable alternative use, may not have been critical at the size to which the British charcoal iron industry grew. But it must have been a potential problem. In a well-wooded country like France, low charcoal supplies in certain areas could limit or even extinguish one regional branch of the industry at a time when landowners in other provinces were avid for the arrival of furnace industries to get some revenue from otherwise worthless woods; even in Sweden strong measures could be imposed in some iron districts to limit the utilisation of woodlands. The critical question, not really answered by historians, is how much further could English ironworks have gone into the countryside and continued to find areas for continuous coppicing, unaffected by the other competitive uses for land, and close enough to ores of sufficient yield and to sources of ample water power. That the late eighteenth-century expansion of the industry could have taken place on a basis of charcoal is inconceivable [25]. However, that the industry extended as far as it did must have been powerfully affected by a factor little mentioned and difficult to assess. This is the fact that the rest of British industry had turned almost entirely to a source of fuel other than wood. If glassmakers, copper smelters, brassmakers, saltboilers, steelmakers, dyers, brewers, sugar boilers and a long list of other industrialists had been dependent on wood for their fuel, ironmasters would have been surrounded by vociferous fellow consumers not only bidding against them but appealing to the state for protection. Nor did the ironmaster have to fear the opposition of the domestic householder. London, with its half-million inhabitants by 1700, had long been dependent on the 'sea-coal' shipped down from the Tyne and Wear, and a high proportion of other domestic consumers both in town and country used mineral fuel – even the workers of Dean, living in woodland essential to the iron industry, often used coal on their fires at home. In Britain all these other consumers had moved to coal because it was cheaper, and the fact that they did so left the ironmaster in a lucky isolation. The sole emphasis in textbooks on the fuel needs of the charcoal iron industry must leave students with the impression that the conversion of the iron industry to coal was a first step in establishing a coal fuel technology in Britain, when it was virtually the last [34; 62, vol. I Part II, *Chs. II, IV*, esp. *245 seq*].

(iii) FAILURE TO SATISFY THE HOME MARKET

The best case for there being no charcoal shortage at the output the charcoal iron industry reached over the first half of the eighteenth century is the apparent steadiness of charcoal prices between the 1690s and the mid 1750s [43, *32 seq*; 32, *609*, gives figures for 1692–1748]. But, even if it was not crippled by its fuel supply, there are reasons for considering the British charcoal iron industry as something of a failure, because through the eighteenth century, until the time when coke iron took over, it could not supply all its national market. This was not a marginal matter; at least half the market was supplied from abroad for a considerable period. England had imported iron in the Middle Ages, mainly from Spain. Swedish iron began to come in during the first half of the seventeenth century and became significant after 1650, when the imports first reached 2000 tons. In 1700 they reached 12,000 tons, by 1720 (a year of war in the Baltic) 14,000 tons, and from then till the mid 1760s they were close to home production of about 20,000 tons of bar. Swedish iron entered so readily because it was highly competitive in price, and that competition had already depressed English prices before 1700 [4, *129–31*; 39, *3 seq*]. However, in years when war or other obstacles impeded Swedish imports, home prices improved and there was a tendency to build new ironworks in Britain. A Swedish author has recently put forward the thesis that the tendency for ironworks to flourish in the West of England, the West Midlands, Wales and the North West may be in some degree caused by the fact that these areas were geographically least affected by competition with Swedish imports which entered at London, Hull or Newcastle [4, *137 seq*]. There may be something in this, but more important in the location of English works was probably the relation between fuel, ores, power and markets. There does, however, seem to be evidence that Wealden, and especially Sussex, works were particularly affected by Swedish competition, and retreated into a considerable concentration on castings for which local ores were particularly suitable [78, *61–3*; 4, *136*]. For steelmaking, however, Swedish iron was essential, and for anchors it was very desirable [39, *21*] so that there were parts of the market where its hold was deservedly strong on qualitative grounds. There were combined efforts by English iron merchants and Swedish diplomats to avert duty on Swedish iron, which however received a 10 shillings per ton duty in 1670, raised further in 1690, 1700 and 1705. It has been calculated that in 1700 Swedish iron faced customs and other dues amounting to more than a quarter of its selling price and leaving a gap of only £1 between the price per ton of English and Swedish iron.

In the years from 1709 to 1720 Swedish iron was adversely affected by war, by their own export duties, by the blockade of Swedish ports, and by a brief English ban on Swedish trade. Particularly in 1717 and 1718 the trade was greatly reduced, English prices rose by 30 per cent and English production boomed [4, *140–1*; 3, *108–18*]. But, despite deliberate Swedish restriction on exports after 1720, our home producers' share of the British market measured in pig iron equivalents fell from 54 per cent in 1716–20 to 43 per cent in 1750, in a period when British consumption rose by about 50 per cent. It is not possible to follow all the fluctuations of the trade; the 1730s were a period when British ironmasters were seriously affected by foreign competition, but the period of the War of the Austrian Succession, 1740–8, seems both to have limited foreign imports and stimulated home production.

The complex issues and arguments in Britain concerning the possibilities of importing pig iron from the North American colonies, while trying to restrain them from producing wrought iron and the many iron manufactures made from it (which would limit our export market for iron goods to them) cannot be pursued here [3, *113–27*]. From the 1750s American pig iron imports were permitted without duty, but they were generally about a tenth of the imports of *bar* iron from other countries and consequently even less after the pig had been converted to bar iron at British forges. From the middle of the century, especially when war or foreign policies reduced Swedish imports, Russian imports of bar gained ground, were much more important than American imports and from the 1760s were generally greater than Swedish [39, *9–15*].

(iv) COMPARATIVE LABOUR COSTS

The reason for the much lower production costs of Swedish and Russian iron is now substantially agreed. The main problem is not seen as a charcoal famine in Britain forcing up the cost of purchasing coppice woods for charcoal production (though wood cost here are likely to have been higher) or even in the poorer supply of water power, or possibly in the poorer quality of ore. It is thought to be due to the higher costs of labour in this country than in either Sweden or Russia. These were much involved in the felling of timber, the work of charcoal burning, the transportation of charcoal to furnace and forge, and in the extraction and transport of ore [23, *151–3*]. Given the higher production costs of British iron, and the fact that certain grades of iron could not be produced here, the British producer was unable to hold more than about half the home market. Here he was protected by the import duties, by transport costs of foreign iron to some parts of the country, and by the fact

28

that British iron was better for some industrial purposes – for instance it dominated the great market for nails and hardware iron. Even the great advances in both the technology and the output of the British industry in the second half of the eighteenth century did not lead to a swift reduction of total imports of wrought iron. These averaged well over 40,000 tons per annum in the last quarter of the century, and did not fall off significantly until after 1800 [39, *10*].

3 The British Technological Revolution

We must now turn to the great technical advances which marked the development of the British iron industry in the eighteenth century. It had been obvious since the later sixteenth century that the possibility existed of lowering the costs of iron production by substituting mineral for wood fuel, a substitution which was taking place throughout fuel-using industry in Britain. Consequently there was a long series of attempts to use coal to smelt iron, which can be only partly indicated by patents. For instance in 1631 a Shropshire landowner advised his heirs, 'It may fall out iron may hereafter be made with pit-coal, then my coal will stand instead for my furnace, and coal may be brought to my furnace by waggons' [81, 7]. Several attempts to make iron with coal had been made, even within the Forest of Dean, by the 1680s [37, 257]. The successful melting of glass ingredients with coal and the total compulsory conversion of the English glass industry to coal use, the important invention of the smelting of copper with coal in a reverberatory furnace at Bristol in the 1680s, and the similar smelting of other non-ferrous metals, made iron the laggard. Progress with coal smelting of iron was disappointing, but this was certainly not because the endeavour was regarded as unimportant, but rather because the sheer technological difficulty was greater than that of converting most other industrial processes to coal use. The fruitless outcome of the celebrated claims of the West Midlands ironmaster Dud Dudley in the 1650s and 1660s – though whether he did have temporary success has been debated – may have discouraged others [54, *17 seq*; 53, *48 seq*].

Iron was first smelted with coal in the form of coke by Abraham Darby I at Coalbrookdale (now part of Ironbridge, Telford) in 1709. This was one of the greatest advances in the history of technology, on which subsequent ferrous metal production in the modern world has been, and still is, based. But its immediate impact was very limited and the circumstances of its introduction are obscure. Darby, born 1678, came from a Quaker locksmith's family of Wren's Nest near Dudley, close to the area of Dud Dudley's operations. His father was a farmer, nailer and locksmith. Abraham was apprenticed to a Birmingham malt mill maker, a trade using cast iron. In 1699 he began to make malt mills in Bristol. There he joined in the formation of the Bristol Wire Co. (or Bristol Co.) which successfully made brass in 1702, and he was involved in their attempts to make brass pots according to the best European technology [66; 81, *13 – 11*]. Joan Day believes his journey to obtain the technology was to the Aachen area and for hammered and not cast pots, as has usually been thought [20]. His part in the Bristol company must have familiarised him with the smelting of copper with coal, in which coke was sometimes employed at that time; as a malt mill maker he must have been familiar with the use of coke in malting. He left the Bristol Co. in 1706 and endeavoured to cast iron pots in a small *foundry* in that town. He succeeded through the efforts of an apprentice and patented his method in 1707. In the following year he took a lease of a recently abandoned and damaged blast furnace at Coalbrookdale, with plentiful coal and ore supplies, adequate water power and the valuable water transport of the adjacent Severn [56; 33]. We now know that Darby did not come to Shropshire with this single object in mind. He was also concerned to establish copper and brass works, and even engage in copper mining, and there is good evidence that he actually was involved in such enterprises for some years. (Information from Dr Barrie Trinder.)

It was formerly suggested that Darby's introduction of coke in iron smelting was gradual and took a few years. There is indeed family testimony of a much later date that he began with charcoal and gradually substituted coke, but a documentary source which the Victorian author Samuel Smiles used and which he claimed substantiated this, can no longer be traced [66, *38* quoting 76, *39*]. Ashton emphasised three things: the essential qualities of the coking coal of the Coalbrookdale district; the desire and ability of Darby to keep his discovery secret; the inferior quality of the coke pig iron produced by him. These he believed accounted for the delay in the spread of the process out of the Ironbridge Gorge before the 1750s [3, *29, 32 seq*]. Here Ashton's views have been countered by the recent writing of Hyde. He strongly doubts if Darby could have kept his secret for a lengthy period, and this is rather unlikely. Darby had partners, persons associated with the firm travelled widely in the country, it was notoriously difficult to prevent workmen leaving if they wished, and the lighter high-quality castings the firm produced must have attracted attention and curiosity as to what was being done at Coalbrookdale. Attempts were made to use coke at a nearby furnace under different ownership almost immediately after Darby's discovery. Hyde's point must be accepted, but even if the smelting secret could not be preserved very long, the close-knit Quaker community at Coalbrookdale would probably be less 'leaky' than most, and Hyde accepts that foundry secrets could have been preserved there for some years [43, *23–5, 41*; 81, *15–17*].

The quality of coal used by Darby and its coking properties were, Hyde agrees, important. The 'clod' coal of the locality was low in sulphur, while that in some other regions contained between twice and six times as much. Excess sulphur could give 'hot shortness'. But he points out that after 1750 ironmasters in many other regions were able to find local coal suitable for coking, while before 1750 other Shropshire ironmasters with ready access to clod coal did not change from charcoal to coke. Again, Hyde does not regard the pig iron of the Dale as being of such a poor quality that it could not convert to wrought iron in the forge, and he rejects Ashton's view on this as reiterated by Flinn and others [24, *67*; 9, *28–9*; 27, *33*; 66, *54, 68*; 72, *331*]. He shows that Coalbrookdale Co. did convert its pig at certain periods; rather it was the *cost* of converting it which was the problem. It was no more affected by sulphur or phosphorus (the latter was the main cause of cold-shortness [55, *81*]) than charcoal iron made with similar ores would have been. Hyde claims that the problem with coke iron was its greater *silicon* content because of the higher furnace temperature with coke, and this remained a constant feature even after the wide spread of the process from 1750. The silicon made it more costly to turn coke iron to wrought iron at the forge, increasing the consumption of charcoal, the loss of pig iron in conversion, and labour costs; it did not mean that the resulting iron would be poorer, only more expensive [43, *27–40*].

Hyde's explanation of the slow take-up of coke iron pig for conversion to wrought is based upon surviving furnace accounts and these have become an important tool of the historian. Mott originally put forward the view that the furnace accounts at Coalbrookdale for 1709 indicated that Darby used coke exclusively there from the beginning [55, 7; 56, *68 seq*], and Hyde seems to accept this [43, *24*, *33*].

This then is the current orthodoxy. Mott concedes that Abraham Darby I at certain periods added a small quantity of charcoal (and possibly occasionally peat) to his coke to make the furnace work better [56, *68–70*; 81, *13–14*], but the older view that Darby gradually converted his furnace fuel from charcoal to coke is discounted by both Mott and Hyde. It had been founded on a nineteenth-century statement by Samuel Smiles, based on his reading of a Coalbrookdale Co. document, a Blast Furnace Memorandum Book, which is now lost, and on the evidence of two eighteenth-century Quaker ladies, admittedly closely connected with the company, but written down long after 1709. Mott ingeniously explains away small payments to a 'wood coll[i]er', or charcoal burner, in 1709 [56, *91*].

(i) THE GENERAL ADOPTION OF COKE IRON

It is accounting evidence which is now used to explain the long period before coke smelting was generally taken up. Abraham Darby I built a second furnace at Coalbrookdale in 1718 which operated successfully; there was experimentation by other owners at a nearby furnace at Kemberton which was not persisted with, a coke furnace at Bersham was unprofitable to its first users, though famous later on; one was started at Willey near Coalbrookdale and there were others at Clifton and Maryport in Cumberland, Chester le Street in Durham and Redbrook in Dean, not all of which were successful [56, *84–8*], and possibly a short-lived one at St Helens. The output of coke furnaces was at first low [81, *16*], and while Darby may have been able to produce very salable castings (particularly thin castings) at a profit, his process would be unattractive to charcoal ironmasters whose aim was to satisfy the much larger market for wrought iron. Hyde stresses that what would induce charcoal ironmasters to change to coke iron furnaces would be a decisive shift in total costs in favour of the coke process: these would include not only variable costs, of which raw material and fuel costs would be the most significant, but the less important (though not negligible) capital costs. He concentrates first on variable costs and compares the costs of 14 charcoal furnaces between 1710 and 1750 with Mott's data derived from the Coalbrookdale accounts for those years in which data survive. Hyde claims that through the 1720s and most of the 1730s the Coalbrook-dale Co. was a higher cost producer, though its furnace operating costs became equal to those of charcoal furnaces around the end of the 1730s. Its capital costs, however, and thus its total costs, remained higher. But even if pig costs at both kinds of furnace had been equal, charcoal pig would have still been preferred by forgemasters because it was cheaper to convert to wrought iron in the forge, so that a decisive cost advantage in favour of coke pig, perhaps of £2 a ton, was needed if the process was to become popular and new coke furnaces were to be built [43, *56 seq*].

Something certainly shifted the balance in favour of the coke iron producer about mid century. Over 1750–71, 27 coke furnaces were started, usually with a much bigger output than the 25 charcoal furnaces which closed in those years. The coke furnaces were often concentrated in groups, for example 5 at the Carron Co. works in Scotland, 6 at those of the Coalbrookdale Co. North and South Wales, especially the latter, also saw new coke furnaces. This expansion clearly connects with a new and massive use of coke pig for conversion to wrought iron at the forges; for instance the new coke furnaces of the Darbys at Horsehay sent 90 per cent of their product to the forges in 1755–61 [43, *55*]. The statement by Abiah Darby (widow of Abraham Darby II) that about 1749 her husband found some way of producing a new quality of coke pig suitable for the forges is rejected by Hyde, who believes that no qualitative change took place, and that all depended on changing fuel costs. 'Charcoal prices rose sharply in the early 1750s in all the major iron making districts and the cause of this increase is not entirely clear, [43, *57*]. It may have been rising labour costs, thinks Hyde, but more likely it was an increase in demand which forced up pig iron prices. An increased demand for iron would result in additional demand on inelastic charcoal supplies, charcoal prices would rise and this would put up the costs of charcoal furnaces. Technology in charcoal iron production was stagnant, so that increased charcoal prices forced up variable costs abruptly. For the coke iron producers the situation was quite the opposite. 'Fuel costs fell more sharply than any other single component of total costs. Ironmasters had greatly improved their efficiency in using coal, while coal prices had fallen, especially relative to charcoal prices' [43, *62*] so that coke iron furnaces saved £1.50 to £2.0 a ton compared with charcoal ones, perhaps as much as £3.0 in the 1760s. Resource supplies, markets and managerial efficiency were not equal at all charcoal iron furnaces, but the fact that over 1750–88 their numbers decreased from over 70 to 26 is strong testimony to the better economies of coke furnaces. The trend continued, so that about 1790 there was an output of 80,000 tons of pig from coke furnaces, while charcoal furnace production was then a mere 10% of the national total.

The adoption of coke smelting has sometimes been linked with the use of the Watt steam engine in conjunction with an iron blowing cylinder (i.e. an air pump) to provide blast, replacing the older leather-and-wooden bellows. But most of the first wave of new coke iron furnaces after 1760 did not initially have steam engines, and only a fraction used Watt engines when they became available after 1775 [43, 69–73]. Increasingly, however, iron works did use steam engines. Sometimes, as at Coalbrookdale, where they pioneered the practice from 1742 [66, 65] they used Newcomen pumping engines for the recycling of their water supply to provide power to bellows. Such engines were sometimes employed so that more than one blast furnace could be blown by the water supply available at a single site. By 1790 steam power was widely employed for blowing, and increasingly by means of the blowing cylinder. It removed the need for seasonal closures of furnaces because of water shortage, thus 'increasing output and lowering capital costs' [43, 71], so that a furnace might perhaps only be inactive for an average of a month a year, and not in every single year as it was simply for repairs. As the steam engine was adopted the need for water power was no longer a constraint when siting iron furnaces: we will see that the influence of the engine soon extended beyond the smelting stage. Steam power also allowed districts with ample ore and coking coal but without strong streams to come into their own. Here the most remarkable example was the Black Country, which quickly began to catch up on its neighbour, Shropshire. Very important was the fact that a coke furnace could take a much heavier charge than one using friable charcoal, but only if a more powerful blast could be produced, as it could with steam power and the blowing cylinder. Ability to take a heavier charge implied an increase in furnace size. Darby's Horsehay furnaces of the early 1750s could each produce 700 tons a year by means of a steam engine which pumped water onto a wheel whose shaft operated the bellows. By the late 1780s coke furnaces averaged around 1000 tons a year, around 1500 tons in 1805. The Neath Abbey works of 1793 had furnaces 60 feet high and a blowing cylinder 70 inches in diameter [44, *22–3*]. Furnaces continued to be built close to ores and to their fuel – now near to coal fields rather than woodlands. But their access to resources was greatly improved by the use of railed roads, first with wooden but increasingly – after their invention in 1767 at Coalbrookdale [81, *72–3*] – with iron rails, and by the great canal building of the late eighteenth century. Railed ways and canals reduced ironmasters' raw materials costs and thus their total production costs, canals and turnpikes also reduced the delivered price of iron to consumers, but calculations of these very important economies are not readily available.

The great impetus to change from wood to coal fuel in the middle of the eighteenth century is thus now seen as mainly a matter of changed fuel prices rather than as a specific technological break-through. But Hyde does admit to some improved technical management of coke furnaces; 'roughly half the cost reduction [in coke-smelting] can be attributed to increased efficency, while the remainder was a result of lower factor prices' [43, *61*]. Mott indicates a progressively-reduced phosphorus content of coke iron (related to an increased iron yield from ore smelted) in the Darby furnaces [56, *80–1*]. Again there is evidence from the widow of the second Abraham Darby that he sent coke iron to forges about 1750 without declaring its origin, to see if it would be accepted by forgemasters and if their prejudices against it could be got over. She also said that the great forgemaster Edward Knight, a principal early consumer, urged her husband to get a patent [3, *251*], and there are picturesque accounts of Darby spending days and nights at the furnace during some trials. Raistrick believes that the story 'probably preserves the record of a real incident' [66, *69*]. Possibly improvements in furnace management over time culminated in final efforts which made important improvements in the costs and quality of Darby's iron about 1750, so that the conversion of coke pig to wrought iron, tried for some years in the 1720s at forges owned by the Darby partners, and then abandoned, could be triumphantly revived, without there having been any single invention capable of patent. The almost headlong development of new furnaces at Horsehay and Ketley by the normally cautious Darby partners, already determined on by 1753 [56, (2), *271 seq*] in the face of scepticism and expectation of their ruin by local industrialists [81, *21–3*] supports those who would still claim some significant technological strides in the Severn Gorge about 1750. The rise of charcoal prices, with its effects on charcoal ironmasters, had hardly begun before the Darbys and their associates had made their decision to make a huge, unprecedented investment. Debate on this point is certainly not yet exhausted [68].

Nevertheless the clear evidence from the 1750s of the rising prices of both the fuel and of the product of charcoal ironmasters is now recognised, and shows that the fuel price factor has been too little emphasised in the past. The rise in charcoal prices is indisputable: whether the fuel cost element in coke smelting more or less simultaneously declined because of a more effective use of fuel in the furnaces or because of a marked fall in coal prices in the 1750s, is harder to determine. It is not clear that a significant and rapid fall in coal prices occurred nationally at this time, nor why it should have done. Improvements in mining technology were important but gradual, and the first canals were constructed only in the late 1750s, and then not in ironmaking districts. But it seems undeniable that some improvement in demand for iron in the 1750s resulted in a situation where charcoal ironmasters could only meet it with increased costs of production and prices of iron, at a time when those of coke ironmasters were falling. Whether they fell because of falling coal prices or because of technological improvement is not quite certain, though the former is the view most recently advocated. In any case for the charcoal iron producers the writing was on the wall.

(ii) COAL IN THE SECOND STAGE OF THE INDIRECT PROCESS

The production of wrought iron could have been only partially liberated from the exigencies of wood and water if the forge stage, at which pig iron was converted to wrought iron, had not been freed as comprehensively as the furnace stage. Wrought iron was still the kind of iron most in demand, even if remarkable pioneering by the Coalbrookdale Company, the Carron Company and the Wilkinsons had been greatly extending the uses of, and demand for, cast iron. The most famous and lasting innovations in the employment of coal in converting pig to wrought iron were those of Henry Cort in the 1780s, but they were preceded by others which had already allowed the conversion of pig to wrought iron with coal fuel.

The trend may be said to have begun with the elimination of charcoal and the use of coal in the second or *chafery* stage of the forge process. This had certainly been practised by the Coalbrookdale Company by the 1730s and was general by 1760. The chafery process, however, was simply a reheating one where no chemical change in the iron took place, so that the substitution of a less pure fuel did not cause problems. The next developments are clearly expounded by Hyde [43, *Ch. V*, *76 seq*], developing pioneering work by Mott [57, *47–50*] and by Morton and Mutton [52, *722–8*]. At the beginning of the century there had been the idea around that coal could be used in a *reverberatory furnace*, in which fuel and ore were effectively separated, and the heat of the coal fire was reflected or reverberated down from the roof of the furnace, so as to smelt iron ore directly to wrought iron. The notable success of coal-fired reverberatory furnaces in smelting non-ferrous metals from the 1680s provided the inspiration. This idea lay behind the notorious ironmaking schemes of William Wood in the early years of the century [80, *97–112*; 24, *55–71*] and the disastrous bubble company founded on them, but technologically this was an impasse. From mid century success in producing convertible coke pig, rising demand, and rising charcoal prices increased inventive interest in changing forges completely to coal. There were nine patents for this between 1761 and Cort's patent of 1783–4. The problem of using coal in the finery was that it contained sulphur which went into the iron. Such wrought iron when heated by smiths to work into the goods would prove too brittle (*red short*). If coke pig instead of charcoal pig was used in the finery it had the further problem of a high silicon content, which had to be eliminated by longer and more costly working. So in addition to the normal job of removing carbon, the forgemaster using coke iron as his raw material and coal as his fuel would have two other elements to remove [43, *83*; 52, *722–8*].

The first practical process was that of the Wood brothers of the early 1760s, developed after long experimentation, perhaps suggestive that their father, though a reckless speculator, had been a serious (if mistaken) technologist. This was the potting and stamping process. First silicon was removed from coke pigs by melting them with coal, but though the iron lost silicon it gained unwanted sulphur from the coal. To deal with this, the iron when cold was broken into small pieces by heavy stamps and put into covered clay crucibles. These pots, up to 20 at a time, were heated in a coal-fired reverberatory furnace, some lime being put into the pots as a flux to absorb the sulphur. The high heat of the furnace oxydised the carbon, the suphur was absorbed into the flux, while the pots prevented further impurities entering from the coal. The pots were made so as to break in the heat of the furnace once their job was done, and the iron contained was then taken to a coal-fired chafery in the ordinary way. Further patents substituted coke for coal in the first refinery heating; this meant that little sulphur was added in removing the silicon and a flux was now no longer needed in the pots. Because mechanical stamps were used to break up the iron so that it could be put in the pots the whole method was called the *'stamping and potting'* process. It was developed by the Wood brothers in Cumbria and in the West Midlands, was widely adopted from the late 1770s and was not a mere brief, very inferior, prelude to Cort's process; indeed until the mid 1790s it converted more coke pig than his process did, and it was the main agent in the elimination of charcoal forges, because this method first undercut their prices [43, *82–5*; 52, *723–5*, 58 *Ch. 1*]. The basic process was developed by the Woods and improved by further patents taken out by Cockshutt and by Wright and Jesson in the early 1770s.

The *puddling and rolling* process of Cort was however the classic complement to coke pig, and once perfected spread nationally and internationally as the best way of converting coke pig to wrought iron. The only significant exception was iron for steelmaking. Its domination lasted until the arrival of the cheap steel processes of Bessemer and Gilchrist-Thomas in the second half of the next century led to the supersession of wrought iron by steel as the main product of ferrous metal production. Even then it continued to provide the raw material of crucible and cementation steel (see below) until they were slowly eliminated by the newer kinds. As long as wrought iron was in demand it was made by puddling and rolling; it finally died as a commercial process at Bolton in 1973.

Like Darby, Cort, the inventor of the second classic ironmaking method of the eighteenth century, was not originally an ironmaster. He seems to have been fairly well-to-do when he married, as his second wife, a woman whose uncle was a major supplier of goods, particularly iron, to the navy. Cort himself became a navy agent, handling such things as paying-off crews. The uncle had interests in a forge at Fontley Mill in Hampshire, which he bequeathed to Cort's wife, and when the concern got into difficulties Cort took it over. A main function of the forge was the recycling of old iron for the navy, and Cort's first breakthrough was a new method of turning the iron into bars by passing it at welding heat through pairs of rolls, operated by water power, in which there were grooves to give the iron a desired profile. There had been earlier, but apparently unsuccessful, patents for such a process. Cort, however, realised that another possible use for such rolls was to employ them to consolidate the grain of the iron and squeeze out slag [58, *28–36*].

His next development had also had its unsuccessful predecessors, for the Cranage brothers (1766) and Peter Onions (1783–4) had fruitlessly tried processes with some similar elements in Shropshire. Cort's second patent was for the use of a coal-fired reverberatory furnace in which pig iron was heated and then, in a melted state, was stirred (*puddled*) by iron bars. The puddling increased the speed at which carbon was removed. Eventually a bubbling occurred, and blue carbon monoxide flames (*puddlers' candles*) were given off. After continued stirring, the iron began to become spongy and was formed into a large ball of iron and then removed from the furnace to have slag expressed and be shaped roughly by the forge hammer, and finally passed at welding heat through grooved rolls, as described in the first patent. The first part of the process, *puddling* as it was later called, demanded strength, manual skill and judgement; *puddlers* became aristocrats of labour and theirs, like those of the men in charge of the rolling process, were among the important new skills created during the industrial revolution [26, *8 seq*; 27, *43 seq*; 35, *177 seq*]. The skill involved meant that there was a considerable learning problem in mastering the Cort system, and the transfer of the technique from Cort's works took time; the new South Wales works, particularly Cyfarthfa, leased to Richard Crawshay in 1787, were the first to make extensive use of it. The reverberatory furnaces needed no power (they were 'air furnaces' not needing bellows), and the application of steam power to operating hammers and driving rolls in the *forge train* finally freed the iron industry from any dependence on water power. A great feature of the Cort process was the remarkable increase in the speed of operations as the work of the rolls reduced that of the hammer, eventually speeding throughput twenty-five times [58, *Ch. 9* and *74*; 43, *88–103*; 9, *33–43*]. Unfortunately, because his partner Jellicoe died insolvent and in debt to the navy, to which he was a paymaster, Cort was rendered bankrupt, lost the use of his patents and was unable to trade. For a few years before his death in 1800 he received a government pension of £200 a year.

4 The Rise of Steel Production

Steel is intermediate in carbon content between wrought iron and cast iron. Attempts to make it cannot have been influenced by this knowledge before the late 1780s, when it was established in a famous paper by leading French scientists (Vandermonde, Berthollet and Monge, 1786). Similarly the carbon difference between pig and wrought iron had not been known to those in Britain who had, entirely empirically, developed successful new ways of converting pig iron to wrought. The effective superiority of empiricism to science at this period can be seen from the fact that the new knowledge of carbon content did not help the French to make good steel, an area in which they made great efforts but continued to be very backward. The English, however, had made great strides in steelmaking from the middle of the seventeenth century and in the middle of the eighteenth century they developed a new steel of great future importance.

Steel was produced in much smaller quantities than wrought iron but this should not cause it to be dismissed as unimportant before the cheap steel era of the late nineteenth century. It was an essential component of the many cutting tools whose quality governed the standard of work which could be produced by craftsmen, and it provided such domestic items as cutlery and razors. The accuracy of watches depended on the homogeneity of watch springs; the quality of chisels, planes, metal-working tools and some agricultural instruments (like scythes), depended on steel edges, almost invariably welded onto iron bodies to save expense. The quality of the cutting surfaces of metal-working tools was essential to the growing engineering industry of the late eighteenth and early nineteenth centuries; engineering was also heavily dependent on steel files of which England had become the leading producer by the mid eighteenth century [2]. Good quality steel tools not only aided the quality but speeded the work of craftsmen; the contribution of hand tools to industrialisation is an important but largely overlooked subject. Steel, however, was very expensive as compared with iron and most other metals, consequently great efforts were made to use it with economy.

There were in Europe a formidable number of methods of making steel of different kinds and qualities, and these have a bewildering set of names; sometimes even the application of the names is inconsistent. Fortunately most of these methods of production were not used in Britain and they can be ignored for our purposes, while an excellent survey of steelmaking in our period is recently available [7]. Little is known about the production of steel in medieval England, though it is certain that much was imported, especially from Westphalia by Hanse merchants. It was of course important for certain arms and armour which were often edged or faced with steel, as were sometimes the critical parts of ploughs. In many cases the surfaces of wrought iron articles were case-hardened, that is they were heated in contact with a carbon source, frequently charcoal, to a point where suffecient carbon had penetrated the surface of the iron to change the exterior of the object to steel. In the sixteenth century, at the time of the introduction of the indirect process in England, there were a number of attempts to produce the so-called 'natural' steel (a decarbonised pig iron) in Sussex relying on the imported skills of German workers, but this technique was not persisted with [7, *15–29*].

(i) CEMENTATION STEEL

The basis of British steel from the early seventeenth century to the Bessemer era was the *cementation* process. The first mention of it occurs in a work published at Prague in 1574. It was known to be in active production in Nuremberg in 1601 and was introduced into England by Elliot and Meysey in 1614. Their first patent of that year mentions the use of a *reverberatory furnace* and the putting of iron and charcoal into pots to be heated together. By 1617 they had modified their patent, still using a reverberatory furnace, but now indicating that coal would be the fuel. In a reverberatory furnace the fuel and the material to be heated or melted were kept separate and flame from the burning fuel was drawn over the material by chimney draught. The furnace roof, which sloped downwards, caused the heat to be reflected down or reverberated onto the metal

SHEFFIELD CEMENTATION STEELWORKS 1842

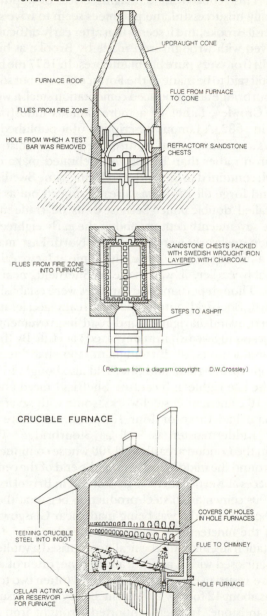

UPDRAUGHT CONE

FURNACE ROOF

FLUE FROM FURNACE
TO CONE

FLUES FROM FIRE ZONE

HOLE FROM WHICH A TEST
BAR WAS REMOVED

REFRACTORY SANDSTONE
CHESTS

SANDSTONE CHESTS PACKED
WITH SWEDISH WROUGHT IRON
LAYERED WITH CHARCOAL

FLUES FROM FIRE ZONE
INTO FURNACE

STEPS TO ASHPIT

(Redrawn from a diagram copyright: D.W.Crossley)

CRUCIBLE FURNACE

COVERS OF HOLES
IN HOLE FURNACES

TEEMING CRUCIBLE
STEEL INTO INGOT

FLUE TO CHIMNEY

HOLE FURNACE

CELLAR ACTING AS
AIR RESERVOIR
FOR FURNACE

CRUCIBLE

43

[28, *170*]. This reduced the damaging effect of impurities in the fuel. Initially unsuccessful, the patentees seem to have soon joined with Sir Basil Brooke, and it seems that after early difficulties success was achieved with wrought iron made by Brooke at his Forest of Dean works from very pure haematite ores. In 1677 the best English steel was still said to be made in the Forest of Dean, but soon the area round Stourbridge also produced cementation steel, a works run by Ambrose Crowley, father of a more famous industrialist, being recorded in 1682. A famous description of it published four years later shows a most important feature, the use of Spanish or Swedish wrought iron rather than British. A celebrated make of wrought iron, which continued to be produced by the same Swedish charcoal furnace and forge till 1941, was already singled out as among the best and called 'double bullet' because of its '00' trade mark. At the end of the seventeenth century and in the early eighteenth several cementation steelworks were built in the North East, mainly on the Derwent, a tributary of the Tyne, and some at Newcastle. By 1692, and probably earlier, it was made at Sheffield, near to cutlery customers. Though perhaps five furnaces were established near or in the town, Sheffield output did not really accelerate until the 1730s when inland navigation and road improvements began to improve access to Swedish iron imported via Hull. By the 1730s the metal manufacturers of Birmingham, too, had steel furnaces, though both Birmingham and Sheffield also bought in North East steel. In the late eighteenth century Sheffield raced ahead and by 1797 had 16 cementation steel works, some with several furnaces. Birmingham had three or four furnaces, and there were then others at Kidderminster as well as Stourbridge. There were furnaces in the London area about 1700 whose continued existence is noted around the middle and even at the end of the century; there were furnaces at Keynsham near Bristol and in Bristol itself in 1725. In the 1790s cementation steel production began at the Cramond works in Scotland, and it was being made near Glasgow by the first decade of the nineteenth century, if not earlier [7, *I*, *Ch. 5*].

Cementation steelmaking classically comprised a vaulted furnace, generally enclosed within an updraught cone, often of a bottle-like profile. Fire-resistant chests were employed (often two to a furnace) of perhaps about 12 foot by 2 foot 6 inches by 2 foot, usually of fire-resistant sandstone. In them imported wrought iron bars made from Swedish Dannemora ore were packed or sandwiched between charcoal so as not to touch each other, up to a total of perhaps 11 tons, and the chests were sealed before the furnace was lit. Coal was the fuel and a temperature of about 1100°C was reached. It took up to thirty-six hours for the chests to become red hot, the heat was kept up for a period of five to nine days and it then took six to ten days to allow the furnace to cool so that the iron could be taken out. If all had gone well it would have absorbed about 1.5 per cent carbon and have become steel, though not a very homogeneous steel. The surface had a blistered appearance, so the product of the cementation process was generally known as *blister* steel.

Once it emerged from the furnace the blister steel was heated again and forged under a heavy water-powered hammer, the steel having been first carefully sorted according to its appearance into different qualities for various purposes and markets, for example '*spring heat*', '*cutlery heat*'. The production of higher grades of steel, progressively improving its homogeneity, was possible by methods probably introduced in the North East about 1700 by Germans. Selected bars of steel of about 2 foot length were bound together into a faggot, heated, forge-welded and then reduced to a bar. This produced the high-grade *shear* steel and, by breaking this into lengths, faggotting them, reheating and reforging, the even more highly esteemed *double-shear* steel could be achieved. As had long been known, extra quality could be given to edge tool steel by *quenching* it hot in water and any excessive hardness resulting could be modified towards toughness and ductility by carefully reheating, or *tempering*. Every file-maker, for instance, performed quenching and tempering operations [7, *I*, *Ch. 3*].

(ii) CRUCIBLE STEEL

In the 1740s a new steel of hitherto unattainable quality was invented in England – up to that time our pioneering in steelmaking had been primarily in the use of coal fuel. All blister steel had been unhomogeneous as shown above, hence the need for selection and reforging. Imperfect homogeneity even in the best steel posed problems for trades which needed a consistent product, such as the watch or clock springmaker. It was a Quaker clockmaker of Doncaster, Benjamin Huntsman, who tackled this problem. Possibly from an analogy with brassfounding he developed the idea that a complete remelting of blister steel might produce a truly consistent product, but for steel a very high temperature was needed, and for this to be successfully attained not only would a specialised furnace have to be devised, but a container which could withstand the heat applied to it and the molten steel within it. Huntsman would have known of the expertise in choosing clays for crucibles and in crucible making for coal furnaces developed by the English glassmakers, and of the use of coke as a fuel in the Darby furnaces. Huntsman moved into the suburbs of Sheffield in the 1740s and there seems to have carried out the significant experiments for his new process, which appears to have been viable by 1750; he was running a steelworks in Sheffield by the next year. The product, *crucible steel* or *cast steel*, was not only suitable for watch springs but also for high-quality edge tools, dies, hammers and small pairs of rolls, invaluable to specialists in small metal goods. It commanded a high price. Huntsman did not patent his steel; it might have been hard at that time for someone dependent on others for his materials (like blister steel and crucible clay), and who also had the final forging done by others, to establish what was critically his. There are stories of industrial espionage, particularly by the Walker family, but all that can be said with certainty is that a number of other steelmakers were soon making crucible steel, among whom the Walkers, who moved from Sheffield to Rotherham, and the Marshalls, were especially celebrated. Huntsman's firm continued making crucible steel until the 1950s though the process was abandoned during the inter-war depression, but finally and briefly resurrected by several firms in the Second World War. Within a short time, English crucible steel, especially that of 'Huntzmann et Marchall' was highly prized on the continent and was described by a Swiss in 1778 as 'the most beautiful steel in commerce'.

The process required the placing of a crucible in a deep bed of incandescent coke, and the furnaces were built over a large cellar whence a powerful air supply went through the furnace, the draft being assisted by tall chimneys in rows, one for each furnace. The narrow furnace or *hole* was originally about 1 foot 6 inches square, formed of refractory sandstone blocks and situated in a working floor which was also the roof of the air supply cellar. Blister steel pieces were put in pre-heated crucibles, originally under a foot high, but soon 1 foot 6 inches, raised to an incandescent heat for 4–5 hours, and then *teemed* (poured) into iron moulds to form ingots.

Good statistics, or even estimates, of eighteenth-century steel production are hard to come by. The Sheffield and Rotherham makers were certainly making well over 2000 tons of cementation steel at the very beginning of the nineteenth century; production in the Birmingham and Newcastle areas was considerable and there was a scatter of other works. At this time about 4000 tons were apparently being exported annually, however, so that, if one assumes half production to be exported, national make would have been upwards of 8000 tons. It must be said that these figures present difficulties which cannot be explored here. There is no sure indication of the amount of cementation steel converted to crucible steel at this period [51, *145*; 7, *I*, *94*; 73, *28*].

There were no further fundamental technical advances in steelmaking in our period. Output figures before 1850 are hard to come by, and generally based on Sheffield, where there was a heavy concentration of production. In 1801, 3000 tons of Swedish iron went there to be converted to steel, in 1835 it was stated that 12,000 tons of blister (cementation) steel were made there; in 1842 (a depressed period) 16,250 tons (and 4000 in the rest of Britain). Half the Sheffield blister production of 1842 was said to be made into crucible steel. By 1843 Sheffield blister production had risen to 21,400 tons, 25,250 tons had been reached in 1846, 40,000 in 1853. By this time some British iron, around 7000 tons, was the basis of part of the blister production [7, *II*, *322 seq*].

5 Iron Production and Markets Before 1800

(i) OUTPUT

In the last dozen years important work on iron production has been published by Hammersley [32], and particularly Hyde [43] and summarised by Riden [69]. The figures are generally arrived at by counting furnaces and forges and trying to find a reasonable average output figure by which to multiply them. Furnaces, being fewer than forges, are easier to count, and the output of pig iron effectively suggests the output of wrought iron, making allowance for the loss of weight in converting pig to wrought (a proportion of 1.35 to 1 is suggested for the early eighteenth century). Of course some iron was cast into goods direct from the furnace or by remelting pig. A figure of only 5 per cent of national production being marketed as cast iron has become ritual for the late seventeenth or early eighteenth century. This may be normally correct but, as has already been noted, in wartime demand for cannon and shot increased the demand for castings. However, with the expansion of coke pig production, which had good casting qualities, and was employed in a greatly increased range of cast goods, the proportion of cast iron marketed had risen substantially by 1800.

National output of pig iron was probably around 23,000 tons in the 1650s, fell to 19,000 tons in the 1670s and recovered to the previous level in the last decade of the century; it climbed a little to 25,000 tons in the second decade of the eighteenth century. While there has been a considerable discussion of contemporary estimates, a 25,000 ton production of pig iron is compatible with that of 18,000 tons of wrought iron which some contemporaries claimed in 1718. That year was one of Baltic war and prohibition of direct imports from Sweden, so there was no reason for depression in the home industry, and the relatively good figure for 1718 helped to provide a useful comparison when British ironmasters wanted to tell a woeful tale based on the bad year of 1737 in order to argue for more duties on Swedish iron. Hyde, however, feels that 23,000 tons of pig would be a safe average figure for the years 1716–20. The 1720s were good on the whole for ironmasters, product prices remained high and (for some unexplained reason) charcoal prices seem to have fallen in the mid 1720s. From the later 1730s the situation worsened, product prices fell faster than production costs. An adverse influence on product prices was competition from increased iron imports in the second half of the 1730s, when they averaged 27,500 tons – greater than they had been in the first half of the decade, or were to be in the 1740s. In the late 1730s propaganda from the industry claimed native output of *wrought* iron was only a little over 12,000 tons, implying a pig iron production of only a little over 16,000 tons; the majority (over 60 per cent) of the iron consumed in Britain was said to have been imported. Hyde does not quarrel with the 1737 statistic, but nevertheless denies a *general* decline between *c*.1720 and 1750. At mid century the number of operational furnaces was roughly the same as in the late 1720s, about 70, despite the net reduction of 9 furnaces in the troubled 1730s. On average the output of the 18 new furnaces built between 1720 and 1749 was much greater than that of the 20 closed over that period, so Hyde would posit a pig iron production of 28,000 tons around 1750. A good contemporary estimate of wrought iron production about this time gives only 18,800 tons, which would consume much less than 28,000 tons of pig – this apparent gap Hyde explains by evidence suggesting that cast iron production might already be 20 per cent of iron production rather than the 5 per cent of the early part of the century (43, *219–20*].

As we have seen, coke iron production rose rapidly during the 1750s and 1760s. To what extent this was due to a favourable shift in fuel cost as compared with makers of charcoal iron, or to what extent due to an improvement in the technology of coke iron producers which effectively reduced their costs, is not wholly clear. By 1770, 40,000 tons of pig were made, and estimates for the late 1780s suggest 30,000 to 35,000 tons of *wrought* iron alone were produced, while cast iron production must have been rising fast as well. An apparently careful estimate of 1791 stated that 85 coke furnaces then made over 80,000 tons of pig, a mere 22 remaining charcoal furnaces produced 9500 tons. Riden revises estimates for 1796 to suggest a 120,000 ton pig iron figure. But 47 new furnaces were built in the five succeeding years. Many were tall, cylinder-blown furnaces and all coke-fired, so that the estimate for 1802 is about 220,000 tons. Since the second decade of the eighteenth century British iron production had multiplied about eleven times, and about eight times since the middle of the century.

(ii) THE PLACE OF IMPORTS

If we are to think of a national iron *market* in the eighteenth century it has already become plain that we cannot merely think of British production. Unfortunately the best estimates of British production are in pig iron and those of most imports are in terms of wrought (bar) iron. Some American pig iron came in after 1730 but before 1750 it averaged less than 2500 tons a year, and even in the 1770s, just before the War of Independence, it averaged well under 4000 tons, and a good deal less in terms of the bar it was converted into in Britain. From the mid 1760s around 1000 tons of American bar iron came in annually. In the 1720s and 1730s around 1700 to 1800 tons of bar a year had come in from Spain [39, *8–11*]. American and Spanish sources are clearly minor compared with those of Sweden and, later, Russia. While there were naturally some fluctuations, the most remarkable thing is the steadiness of Swedish imports from the 1720s to the 1790s. As total imports tended to grow until the mid 1790s, this means the Swedish proportion fell. Over 1720–99 an average of about 17,500 tons of bar iron came in from Sweden, the highest imports coming in the 1740s, 1760s, and 1790s. The relatively static level from the 1720s was very much one deriving from a conscious Swedish policy of trading on the premium position Swedish iron was thought to enjoy in foreign and specially the British markets, and of maintaining prices by restricting supply. In fact this did not work particularly well because of the expansion of Russian exports. Swedish imports into Britain were 75 per cent of the total over 1720–50, 66 per cent in the 1750s and 56 per cent as late as the early 1760s. But in the late 1760s they were only 43.5 per cent; from 1770 to 1799 the proportion fluctuated between 30 per cent and 42 per cent. From the late 1720s Russian iron became significant and imports steadily grew thereafter, they were half as great as Swedish imports in the early 1750s, and 38 per cent of total imports in the early 1760s. Russian iron took the lead in 1765 and virtually retained it for the rest of the century. Peak years of Russian imports, at well over 30,000 tons, were greater than the highest yearly eighteenth-century imports from Sweden [39, *10–15*].

While there may be hidden traps in putting together figures compiled at different periods by different scholars for different purposes one may estimate the British market in the 1720s as averaging around 28,000 tons of home pig output plus 19,650 tons of imported bar, equivalent to about 25,500 tons of pig, so that, in terms of total demand for iron, foreign imports nearly equalled home production. In the early 1750s British annual pig production was still only about 28,000 tons, of which 2000 tons were the new coke pig. Foreign imports averaged over 29,000 of bar in that decade, the equivalent of nearly 38,000 tons of pig, giving a rough total demand of 66,000 tons expressed as pig. The importance of the transformation of the industry after 1750 can be seen from the figures at the end of the century. In 1799 British pig production alone was 170,000 tons, and average bar imports for the last decade were 45,500 tons or about 58,500 pig equivalent, so that, by the end of the century, the formerly dominant position of imports had been greatly changed. After reaching a peak of over 50,000 tons of bar in the early 1790s they had begun a distinct fall in the second half of the decade which had become dramatic by 1815.

(NOTE: These figures cannot be exact; we know for instance that there were changes in the amount of pig needed to produce a given amount of bar over the course of the century. So the multiplier used to convert bar iron importation figures to pig equivalents should ideally be adjusted somewhat over that time, but the sources do not allow it to be done.)

(iii) THE PLACE OF EXPORTS

There is an apparent paradox in terms of British overseas trade in iron in the eighteenth century. A country which was overwhelmingly an importer of iron, sometimes to the extent of importing more than its total home production, was also in fact a considerable exporter of iron. There are a number of explanations for this. We retained for ourselves the iron trade with our colonies, which explains the considerable export to the West Indies and, before the American Revolution, to the North American colonies. Again, the colonies were often interested not so much in pig or bar iron – the northern colonies being in a position to export such iron to us – but in goods made of iron. For their manufacture there was abundant skill in Britain, while some of the iron produced in Britain was particularly well adapted for hardware products and for nails. Swedish iron was converted to steel by our highly efficient steelmaking industry, and some exported in that form. But much depended on our large and cheap supplies of coal which, even when most wrought iron was still itself made with charcoal, was the invariable fuel of smiths of all kinds in making that iron into goods. Coal also enabled the supply of an increasing market in coke iron castings, whether cast directly from the furnace or from pig remelted with coal. Cheap coal and a suitable home-produced iron were the main influences behind the great export of the nails made in thousands of little forges by nailmaking families. The most famous concentration of these was in what was later called the 'Black Country' of South Staffordshire and parts of Worcestershire, but South Lancashire and the Sheffield area were also considerable producers of nails.

Some tonnage figures illustrate the rise of these exports in the eighteenth century; but the terminology of the Customs must be understood; their 'iron wrought' to them meant worked-up or manufactured iron, not the exportation of wrought iron in bar form, which was trivial before the 1790s.

	Nails (tons)	Wrought iron (tons)
1700–9	500	1,020
1750–9	1,600	6,190
1790–1	1,370	17,440

[39, *19*].

The 'official values' of iron exports (which were not regularly revised by Customs after the early years of the eighteenth century and are therefore simply to be used as an index) stood at £85,000 in 1700 (at a period of some fluctuation), passed £100,000 in 1714, £200,000 in 1736 and £400,000 in 1748. Around this level they fluctuated for some years and did not pass the £500,000 mark until 1760. Just before the American War of Independence, which particularly depressed the iron trade, they were commonly around £700,000, but nearly touched £900,000 in the exceptional year of

52

1771. This may be put against total exports of English produce and manufactures which rose from £3.7m in 1700 to £9.7m in 1775. Iron exports passed the £1m mark in 1791, and fluctuated between a little below the million and a little under £1.5m for more than a decade (from 1792 the figures cover Great Britain, Scotland and Ireland then being included). In 1808, the last year of a celebrated series of overseas trade statistics compiled by Mrs Schumpeter, iron exports at official values were approximately £1.2m, as against total national domestic exports of £37m. However, the relative importance of iron exports declined in the later stages of the Napoleonic wars [51, *294–5*; 73].

6 Locational Changes and War

(i) LOCATIONAL CHANGE

The last decades of the eighteenth century and the war years down to 1815 saw considerable shifts in the location of works and in the geographical balance between the major regional centres of the industry. The impetus was largely the structuring of the industry as it concentrated on areas where coal and ore were in proximity, an impetus which became more powerful with the successful adoption of puddling and rolling, removing any need for the dispersal of the forge stage to sites where charcoal and water power were obtainable, once these were replaced by coal and steam power. As might be expected, the first area to expand massively once coke pig could readily be turned to wrought iron was Shropshire, the big push being in the 1750s when five new works were built within seven miles, comprising nine new furnaces. The output of furnaces continued to rise and those rebuilt or newly built were of increasing size, so that while average production of Shropshire furnaces was only around 25 tons a week in the early 1790s, furnaces capable of 40 or even 50 tons were operating a decade later. Shropshire's brief national dominance peaked in the 1770s when it made 40 per cent of national output. Even though Shropshire output continued to grow thereafter, its 33,000 tons output of 1796 had already been passed by that of South Wales with 34,000 tons [81, *345, 136–7*].

The rise of South Wales was spectacular, particularly with the building of works around Merthyr Tydfil between 1759 and 1765 (*Glamorgan County History*, V, 1980, *Ch. III*). The Dowlais works was founded in 1759 by local mineral owners with a large element of Bristol capital. A key figure, however, was Isaac Wilkinson, father of more famous ironmaster sons. He had formerly been a skilled worker in Furness and an ironmaster at Bersham in North Wales, where his sons continued [15, *23 seq*]. Isaac Wilkinson and John Guest, one of the first of the many Shropshire technologist immigrants into South Wales, shortly afterwards founded another works close by at a spot confusingly called Plymouth, in 1763, while in 1765 two men of Whitehaven origin started Cyfarthfa works. One of them, Anthony Bacon, had prospered as a London merchant and government contractor [59]. The Homfrays from Shropshire set up Penydarren works in 1784. Bacon extended his interest by greatly developing the Cyfarthfa works, buying up Plymouth and buying land and building a furnace at Hirwaun. Famous connections were formed when Richard Crawshay, a London iron merchant, took an interest in Cyfarthfa in 1774 and when Guest began to manage Dowlais in 1767. It was Crayshaw who saw through the very difficult period at Cyfarthfa during which the development work was done on Cort's process, final success being achieved by 1791, by which time it had become more economical than the stamping and potting process. In 1812 the Merthyr works together produced one-fifth of British output; in 1815 South Wales as a whole produced 35 per cent, by which time Shropshire's share had sunk to about 12.5 per cent. The other great success story was that of Staffordshire [29, *Ch. 4*; 9, *148 seq*]. There the supplies of good coking coal, ore, lime and fireclay in close proximity came into their own once steam power made good a local deficiency of water power. The canal system supplied cheap transport, and the new technologies allowed the employment of coal fuel throughout the production process. Producing only 2000 tons in 1720, the county produced 125,000 tons in 1815, so that South Wales and Staffordshire then together accounted for two-thirds of national pig iron production and three-quarters of bar. These were the iron districts of large furnaces, some South Wales furnaces producing over 2000 tons a year in the first decade of the nineteenth century, and they were the places where the techniques of puddling and rolling were most developed; Cyfarthfa had the massive rolling capacity of 13,000 tons a year. Naturally capital investment had to increase to sustain such production; about seven South Wales works had capital values of over £100,000 by 1815.

(ii) THE IRON INDUSTRY AND WAR

Even when the iron industry had been much less strong and technically efficient wartime had generally been a period of expanded output and increased profit, especially for those works which were capable of casting guns and shot, though they ran the risk of the rigorous proof that guns had to undergo. During the second half of the century the development of coke iron and of new boring techniques saw the decline of Sussex and the rise of gunfounding in Scotland, the West Midlands, North Wales and Yorkshire. But the French wars between 1793 and 1815 saw a particularly strong demand, while high tariffs on imports of foreign iron for fiscal purposes and wartime interruptions of its supply further aided the British ironmaster. During the war the government was buying 17–19 per cent of total iron production in some years and this was clearly of great importance, though it is hard to speculate what demand from civilian markets at home and abroad might have been had not war disrupted it. At any rate the ironmasters expanded output rapidly and prospered. Costs rose, but not in real terms, because they went up by less than the current inflation; ironmasters' material costs were often held down because they integrated backwards and efficiently into coal and iron mining. On the smelting side there were no fundamental technological advances, but new furnaces were built bigger, and all were driven harder and more continuously. The successful development of puddling and rolling reduced the costs of conversion to wrought iron, so British bar iron became cheaper than imported Russian or Swedish, with a consequent strong decline in their market in Britain. Swedish iron was the less affected because it was still essential as a material for our steelmakers. During the course of the war the British iron industry, long at the mercy of Swedish and Russian importers, and a much smaller producer during most of the eighteenth century than France, emerged clearly as the dominant producer in Europe [43, *Ch. 6*].

The end of the war was very painful for many ironmasters, the year 1816 being particularly bad. The price of pig for conversion to bar iron fell drastically, from £6.0s to £3.15s. In that year production probably dropped to 260,000 tons from 395,000 in 1815, but it then began to recover and had achieved 400,000 tons in 1820. The impact of the abrupt recession was not even; Shropshire, the home of so many of the eighteenth-century advances, had many dilapidated and obsolescent works in 1815 and there were many closures there, though not all permanent. The restoration of production by 1820, and the further advances that came after 1823 [43, *241*; 9, *124*; 69, *455*] showed that civilian had successfully replaced military demand. The main influence here was a great extension of the uses for iron, in which the employment of cast iron was particularly important. While the *foundries* in which goods were cast were originally closely adjacent to furnaces they were increasingly built at sites far from primary iron production. Provided that pig iron could be transported economically and coal obtained cheaply, a foundry could be profitable, and they became a regular feature of the growing number of engineering works in the industrial districts. In addition to the celebrated cast pots of Darby, and flat irons and similar household utensils, the steam engine, of which at least 2500 had been built by 1800 [48], proliferated further after the end of the Boulton and Watt patent in 1800. Cylinders, flywheels, engine beds and, increasingly, beams, were of cast iron, while boilers and some other parts were of wrought iron. Private railways for industrial purposes proliferated before the arrival of the public railway, and the partial use of iron for their rails had begun at Coalbrookdale in the 1760s. The introduction of the L-shaped iron rail for these industrial lines on the *plateway* method came in the 1790s. A remarkable demonstration of the possibilities of cast iron as a constructional material came in 1779, when a group of Shropshire men pioneered the iron bridge across the Severn near Coalbrook-dale, now perhaps the world's most famous industrial monument, and thus foreshadowed further developments including the iron suspension bridge. A Shropshire iron works produced the first iron canal aqueduct, erected by Telford in 1796. Cast iron entered building construction for pillars in the 1770s and the first iron-framed building (still surviving) was built for a linen mill at Shrewsbury in 1796 [81, *Chs VIII* and *X*; *72–4, 123–4, 140–1, 87–8*]. The cotton manufacturer William Strutt further developed this construction method for low fire-risk textile factories. The cheapness of cast iron allowed the manufacture of water pipes and, by the early nineteenth century, of gas pipes. Apart from the steam engine, cast iron developed uses in engineering for gears (replacing wood as the traditional material of the millwright) and the beds for lathes, planers and other equipment, while in the early nineteenth century iron drove out wood in textile machinery, giving increased strength and rigidity (e.g. in the power loom and mule). Though for things like nails, locks, hinges, many agricultural implements and for architectural uses, wrought iron was also made in increasing quantity, cast iron took an increased share of the market. Perhaps 160,000 tons of the 400,000 tons output of the iron industry went into castings in 1815, 100,000 tons for civilian uses, whereas in the early eighteenth century castings had been perhaps 5 per cent of national iron production [43, *127–8*].

7   The Early Nineteenth Century

The first decades of the nineteenth century saw little fundamentally new technology entering the industry. While this was particularly true on the smelting side, nevertheless the general size of blast furnaces increased as coke was taken up, for it was found possible to increase the weight of the charge beyond that of charcoal furnaces. Average output for furnaces in the Black Country, for instance, had risen from about 30 tons a week at the end of the eighteenth century to 40–50 tons in the 1820s, and some made much more. Indeed, powered loading ramps were developed to fuel the higher furnaces. These furnaces were reaching 60 feet by the 1830s, and they were becoming circular in outward profile and internal section, being constructed externally of ordinary brick and inside of firebrick. Furnaces also began to be strengthened by binding with wrought iron hoops; blast pressure was rising and was made more even by regulators (though their value is disputable); internal capacity increased rapidly. By the late 1830s a Black Country furnace produced 236 tons a week [29, *Ch. 5*]. An important process of development was thus going on in smelting even if there was no essentially new technology. In the refining process, however, now universally done by puddling and rolling, there was some significant technical development and a drop in costs; fuel consumption halved and the consumption of pig iron per ton of wrought iron produced also fell. Among the improvements which can be identified are the use of iron oxide bottoms instead of sand bottoms in puddling furnaces, while the furnaces were sometimes charged with molten instead of cold pig. Particularly important was the introduction of *wet puddling* by Hall in the Black Country after long experiments culminating in about 1830 (29, *66 seq*]. He lined the puddling hearths with calcined puddling furnace slag which produced a violent *pig boiling* reaction. This not only speeded the process but helped reduce phosphorus and produced a higher yield of wrought iron from pig.

But the main technical advance of the last part of our period was applicable to smelting, though it did not immediately spread through British ironworks after it was patented in 1828. This was the *hot blast* of the Scottish inventor, James Neilson. The new process overthrew the old belief that the colder the blast the better. Heating the air blown in increased the combustion temperature, helped remove sulphur, and improved output. It also reduced fuel consumption. It led to a great expansion of Scottish production, hitherto minor within Britain, in conjunction with the use of local *black band* ironstone and local raw *splint* coal. Hyde discounts the view that Scottish iron production was under grave threat because of uncompetitive prices before Neilson's process arrived; it was more expensive than Welsh, but so was that of other regions. It was equally protected in its own regional markets because of the cost of transport from producers in other districts, and it was indeed able, before hot blast, to export marginally to some English markets accessible by sea. Hyde also points out that black band ironstone was already used before Neilson's invention, and that where the costs of hot blast furnaces have been analysed the great saving of the process was in fuel costs. This was particularly important to the Scottish ironmasters whose local coal was low in carbon content. They found it possible with the higher temperatures of hot blast to use this local coal direct, uncoked, in the blast furnace, and moved from being relatively high-cost producers to the lowest in Britain by the early 1840s, with variable costs as low as £1.75 per ton as compared with an average of over £3.

The hot blast changed Scotland's relative position in the industry remarkably. The industry there had been small before coke iron arrived, with the celebrated founding of the Carron Company in 1759. For 20 years this had remained the only large ironworks; the first coke furnaces in the West of Scotland were built only in the 1780s. By 1800 Scotland had 10 works, 7 in West Scotland, with a total output of nearly 23,000 tons, 9 per cent of the British total. But thereafter progress was slow, no important new works were built, and only 25,000 tons of pig were made there in 1825, a mere 5 per cent of British output. With the hot blast there was transformation. By 1830 Scottish output approached 40,000 tons, by 1840 it was just short of a quarter of a million tons, 11 new works being built in West

Scotland between 1825 and 1841. By 1848 Lanarkshire alone had 15 ironworks with 92 furnaces built or building and Scotland was producing over half a million tons of pig which had increased to a million by 1860 [11; 75].

English and Welsh ironmasters took up the hot blast, but not at such a headlong pace. Further developments such as the improvement of stoves for heating the blast (1832) and the water cooling of the tuyeres (ends of the blast pipes) to save their excessive heating and wear, helped the progress. Though 55 per cent of iron, including nearly all the greatly expanded Scottish production, was produced with the hot blast by 1840, only 40 per cent of British furnaces were converted to the process. However, almost all works used it by 1860. The expiry of Neilson's patent in 1842 reduced costs and the adoption of the process accelerated. South Wales, already a low-cost producer, was understandably rather slow to adopt the process. In Staffordshire there were doubts about the quality of iron made by hot blast. Once these were removed one gain of the process was seen to be an ability to use previously unacceptable grades of coal to make adequate coke for the furnaces. Hyde's view is that the degree of cost advantage of hot blast varied inversely with the carbon content of the coal of the different iron making regions. While there were other minor improvements, the hot blast was probably the only new technique improving smelting productivity, and thereby reducing costs, before the mid 1840s, and remained the main influence till 1870. There was an almost continuous drop in pig iron prices between the mid 1830s and 1850s, from nearly £7 per ton to £2, with just a short upward rise in the early 1840s, which only reached £3 a ton. The new technology was assisted by a fall of input prices of perhaps 10 per cent between the mid 1840s and the late 1850s, by which time productivity had ceased to rise much. The lower prices before 1850 were influenced by intense competition as well as by technology, and this despite a great rise in demand. British pig production passed the million ton mark in the mid 1830s, it had reached 2 million by 1847 [43, *150–65*].

(ii) IRON AND RAILWAYS

The coming of the railway provided a market of great importance for iron products, but there has been an intricate debate about its extent. Its significance for this study is limited by our terminal date of 1850; obviously the railway demand continued to be considerable afterwards. The subject has also already been expertly treated in this series by Gourvish [31], and there is little to add. However, the most extreme impact of railway demand on the home market for iron occurred in the late 1840s, so the topic should be outlined here. Of course in Britain industry was far advanced before railways came, and there can be no justification for arguing that they provided the decisive push to industrialisation; an iron industry with an internationally unexampled annual output of around a million tons already existed before they became a large source of demand upon it. In the middle and late 1840s about 18 per cent of pig iron output was converted to wrought iron for the permanent way demands of railway companies in the UK, constituting about 30 per cent of the home market, and in 1848 the two figures were 30 and 40 per cent [50; 38, *213*]. But this was quite exceptional, and demand for rails was relatively much lower afterwards. We have to remember nevertheless that the railways required iron for engines and rolling stock and for a number of other purposes, as well as for permanent way. But the demands of the railways cannot be shown to have promoted major new technology in the iron industry; the new market was able to cash in on the hot blast, whose introduction was very timely.

Fast-growing exports, however, contributed largely to iron industry expansion in the 1840s and after, and railway iron, particularly rails, was a major component in those exports. There has recently been much statistically-based discussion about the economic impact of the railways. But many of the figures advanced by proponents of the 'new' economic history depend on assumptions which can only be approximate; the calculation of home demand for rail depends on estimates of rail life; the importance of foreign demand for railway materials depends on the supposition that all bar iron exports were exports of wrought iron rails. If railway demand upon the iron industry was not reflected in the development of new technology, the increased size of the market was apparently reflected in increasing sizes of furnaces and of works, and this would have given economies of scale, though precise measurement of these has not been attempted. Hawke believes that the great rolling-mill demands of rail manufacture did little to influence rolling mill technology or costs [38, *242*]. Hyde points out, however, that bar iron prices declined generally after 1845, but whether this was due to improved efficiency or lower material costs seems not to be clearly established. In some degree the effect of railways on the economy was paradoxical; if excessive railway speculation was a contributor to the collapse of the investment market in 1847, the lagged demand for railway iron deriving from earlier investment, and peaking so strongly in 1848, certainly had a counter-cyclical effect in reducing the crisis which succeeded the 1847 collapse. It may be suggested that the tendency of historians in the late 1960s and early 1970s to mark down the influence of railways on the iron and steel industries (steel is of course really significant only after the end of our period) was rather overdone. In the case of Hawke there is some concentration on rail demand, or permanent way demand, rather than total railway demand [63, *61–2*]. Again the fact that the whole technology of railways was invented in Britain created a demand for railway iron which transcended that of our domestic economy, and it was importantly reflected in exports. There are also imponderables; it is hardly possible to estimate all the backward linkages of railway demand to the engineering industry.

The demands of the railways on the iron industry were not evenly spread through the iron making districts. The Scottish iron industry supplied few rails before the steel age, but may have gained from exports of pig (and other iron goods) to markets south of the border because other ironmaking districts in Wales and England were so occupied with railway orders. The North East of England largely satisfied the railways in its own area, and the Midlands were similarly originally fairly self-sufficient. But South Wales was pre-eminent, especially as a supplier of rails to many other districts, from an early period of railway building. So while the Birmingham Committee of the London and Birmingham Railway, charged with building at that end of the line, largely used Midland suppliers, the London Committee bought from South Wales. Railways in southern England, and the Great Western, also bought from South Wales. By the 1840s the South Wales ironmasters were supplying Lancashire railways and even making inroads into the Midlands and Yorkshire. Their large units, their early ascendancy in puddling and rolling, and their low fuel costs gave them advantages where high transport costs to market did not erode them.

The end of our period crosses a period of major export growth, as iron exports grew nine times between 1830 and 1870 [43, *172*] and eventually they took over half of the pig made, though of course that pig was generally converted to wrought iron. Already in the second half of the 1840s exports had more than doubled, and the largest part of the increase was in railway iron. Estimates of the effect on pig iron production have to be a little approximate, as of course in producing bar iron there was a loss of weight of the pig iron from which it was made. The types of iron in exports also changed. There was a growing export of cast iron goods, machine

Iron and Steel Exports – United Kingdom 1830–1850
(Thousands of tons)

	Pig and puddled iron	Iron in bars or unwrought and railroad iron and steel	Unwrought steel	Total iron and steel
1830	12	68	1	118
1831	12	70	1	124
1832	18	81	1	148
1833	23	83	2	163
1834	22	80	2	158
1835	33	108	3	199
1836	34	98	3	192
1837	44	96	2	194
1838	49	142	3	257
1839	43	136	4	248
1840	50	145	3	269
1841	86	189	4	360
1842	94	191	3	369
1843	155	199	3	450
1844	100	250	5	459
1845	77	164	7	351
1846	159	158	8	433
1847	176	228	10	550
1848	176	339	7	626
1849	162	402	8	709
1850	142	469	11	783

Source: [51].

parts and so on, in which there was not a similar pig loss, and some iron was exported as pig, so that there is some difficulty in arriving at an accurate measurement of exports in terms of their pig content, though this is the best general measurement of their demand upon the iron industry.

There would clearly be a serious gap if nothing were to be said about the ironmasters. The problems of giving a short but accurate account, however, are almost insuperable. The variety of the ironmasters' origins, the complexity of their businesses, the extent of change over our period, and for some areas, the lack of research into the group as a whole (however much effort may have been devoted to the lives and businesses of some very prominent men) make generalisation difficult. All that can be done is to make some broad statements, and quote some examples in the belief that they outweigh the contrary instances which could be raised. It will not be feasible, without resorting to a mere catalogue, to refer to all leading ironmasters, even in passing.

The ironmasters, as Crouzet has pointed out, can be regarded as 'the first group of professional industrialists to emerge', and that group could be recognised even by the beginning of the eighteenth century. Their social origins were diverse. There were still among them men of landed origin. The Foleys and the Crowleys, however, rose from the ranks of dealers and small manufacturers in the West Midlands (while the latter family were not always strictly speaking primary manufacturers of iron, they were over a long period *steel*makers). Others like the Spencers and some of their Yorkshire associates had gentry origins, as did some prominent Furness ironmasters like the Rowlinsons and Machells [19; 69]. The famous Darby family was one of those which originated in the small metal trades of the West Midlands – the environs of Dudley and Stourbridge produced some great iron dynasties. Despite the limited success of the British industry – unable to satisfy much of its own home market in the late seventeenth century and much of the eighteenth – we have seen that some of the ironmasters, like the Foleys and the Spencers, were able to set up sophisticated interlocking partnerships, dominated by the capital of a few partners, to control effectively the output of works remarkably scattered geographically, to handle very large capitals, to serve varied markets and to set up effective (if sometimes idiosyncratic) accounting systems not only for individual furnaces but for the different elements within their grand overall business groupings. Whatever the general and limiting constraints upon the extent of the industry at the opening of the eighteenth century, the ability of ironmasters to build up complex organisations, trade amazingly widely, and occasionally build up great fortunes, showed that entrepreneurial dynamism was already there, and would expand and flourish once technological change and favourable shifts in costs and markets gave opportunity. The original capital of that part of the Foley empire called 'The ironworks in partnership' was over £36,000 in 1692, but as early as 1669 one of the Foleys seems to have had a capital in 'stock and debts' of nearly £70,000. It is interesting however, that both the Foleys and Spencers lost dynamism at the end of the charcoal iron era, and moved from an industrial to an aristocratic or gentry lifestyle [19].

During the eighteenth and early nineteenth centuries there continued to be some aristocratic or gentry entrants to the business of ironmaking, nearly always because they were concerned with developing the coal or iron resources of estates. The Lindsay family, Earls of Balcarres, developed ironworks on the Wigan coalfield in the 1780s, but one influential member of the family was a 'nabob' from the East Indies; other members fitted the role of industrialist less well. They are sometimes held to have been financially unsuccessful in the direct operation of ironworks, from which they appear to have withdrawn before 1815, but they were certainly important in the formation of the great Wigan Coal and Iron Company in the second half of the century. Earl Gower formed a partnership to exploit the minerals of the Lilleshall estate in Shropshire, which operated ironworks after 1792. While his elder son did not continue the interest, his second son was prominent in the Lilleshall Company: this replaced the partnership and became a large concern, and he was also involved in Staffordshire ironmaking. The Earls Fitzwilliam operated ironworks on the Wentworth coalmines in the 1820s and 1830s, but eventually lost money and went back to leasing their minerals to others. The Sneyds at Keele in Staffordshire operated ironworks in the first half of the nineteenth century, though not profitably. But there is no doubt that the aristocracy and wealthier landed gentry were a tiny minority of ironmasters in the eighteenth and nineteenth centuries.

Other entrants came from those involved in the trade in iron. Men called *ironmongers* in the late seventeenth and eighteenth centuries, were not mere retailers but dealers in iron, sometimes considerable merchants in the inland trade. Some put out iron for manufacture, particularly for nailmaking, some of them had interests in forges or slitting mills. The permutations are many. The Crowleys and Darbys came, as we have seen, from the Stourbridge–Dudley district, having risen from the manufacture of small iron goods. The first Abraham Darby had a period of involvement in malt-mill manufacture and the Bristol copper and brass industry before moving into ironmaking at the Coalbrookdale furnace, and for a time he was involved in major schemes to develop the copper and brass industries in Shropshire. The Crowleys moved from nailmaking to being ironmongers, steelmakers and investors in iron forges and furnaces, before the rise of the famous (Sir) Ambrose Crowley III (1637–1713) with his great naval contracts for iron goods and his merchanting, his City interests, and his London warehouses. Around Winlaton and Swalwell on Tyneside he had huge establishments making iron nails and a great range of other items in iron and steel, while continuing some business interests in the West Midlands. It seems that the family ceased to operate blast furnaces shortly after 1725, and in that sense they were no longer ironmasters; they were until the nineteenth century operators of steel furnaces and slitting mills, plating mills and similar plant. Imposing the detailed regulations of a *Law Book* on those who ran his North-Eastern works, and carrying on a ceaseless correspondence of instructions and scoldings, Sir Ambrose showed that it was possible to build up a vast dispersed manufacturing and merchanting business, almost certainly the greatest industrial enterprise of his time. His North-Eastern factories were heavily based on imported Swedish iron, but the whole development showed that, even though currently insuperable forces constricted pig iron output and so the size of the British iron industry, there was already entrepreneurship of remarkable drive and vision about which would take advantage of the opportunity to expand whenever it presented itself. The Crowleys were followed by many later examples of men who moved backward from the production of metal goods in workshops or foundry into the primary production of iron or steel, for instance the partners in Newton, Chambers and Co. of Sheffield in the 1790s.

Merchants, particularly those who either traded in iron or had government contracts, were important entrants to iron production. The two most celebrated were Anthony Bacon, who did so much to launch the South Wales industry and founded some subsequently famous works, and Richard Crawshay. Crawshay was a London merchant who joined in during the later stages of Bacon's operations in South Wales, and whose family subsequently became inseparably identified with the great works of Cyfarthfa. Quaker merchants, such as Thomas Goldney and Samuel Milner, bought or took interests in iron furnaces; indeed the Coalbrookdale Company was for a period sustained and financially dominated by Goldney [66, *12–13*]. Of course many major merchants at this period carried on some informal banking activities, which makes it difficult to be precise about the relationship between ironmaking and banking; perhaps the most striking involvement of banking finance in any large ironworks was that of Scottish banks with Carron Co., which after being sustained by them, was nearly ruined by a banking crisis [12, *21 seq*; 13, *Ch. V*]. The important Butterley Company of Derbyshire, founded late in the eighteenth century, had a banker among its partners. There was one famous instance of the opposite development, the Quaker Lloyds of Birmingham, who had widespread interests in the iron trade, including forges and rolling mills and steelworks, were in 1765 founding partners in the bank which still bears their name.

What cannot be denied is that Nonconformists, and especially Quakers, were remarkably important in the iron industry. In the eighteenth century the Quakers were strongly cohesive, their beliefs stood in the way of integration with the establishment (the move of the third Ambrose Crowley from Quakerism to High Toryism and a knighthood is very untypical) and the obligatory intermarriage within the community strengthened this cohesion. Consequently the Lloyds and the Darbys, at times when there were financial problems or there was no grown son to manage their concerns, were able to call on co-religionists, often relatives, to carry a business through trying times. The Quaker abhorrence of dangerous speculation, the censure heaped on Friends who overtraded and ran into illiquidity, insolvency or bankruptcy (the last sometimes led to their temporary exclusion from the Society) their plain dress, thrift and relatively abstemious lifestyle, meant that they were respected and trusted in trade even by those members of the

69

business community who did not share their beliefs. Quakers and other Nonconformists tended to secure for their children education in private schools or in dissenting academies, where there was greater emphasis on mathematics, accounts or modern languages, subjects which were a good foundation for a life in business. In the early eighteenth century Quakers were said to have owned or operated half the ironworks functioning in England. The most famous ironmaster dynasty of the eighteenth century was that of the Darbys – though its interest in the iron trade did not end with the century, but has continued in one form or other to the present – and Coalbrookdale and the neighbouring works which they and their relatives the Reynolds controlled became a centre of international pilgrimage by foreign technologists and by general tourists, British and foreign. They marvelled at the concentration of industry around the picturesque Severn Gorge, the iron bridge, the furnaces blazing by night, the iron railed ways, the inclined planes linking land and water transport systems, the many steam engines working at coal mines and ironworks, the litter of vast castings of engine cylinders, water pipes and huge iron pots. They recorded their impressions in diaries, while artists filled sketch books and canvasses. Of course with the great expansion of the late-eighteenth-century iron industry the Quaker influence relatively declined, even in East Shropshire, let alone nationally [81; 82, *87 seq*; 77].

Of all the ironmasters of the late eighteenth century the one to make the greatest impact was John Wilkinson, the son of Isaac Wilkinson, a potfounder from a North Lancashire ironworks who had moved into the ranks of the ironmasters in North and South Wales, but without achieving commercial success. His sons John and William were more influential; John was both technically and commercially outstanding. He improved the method of boring cannon and by a separate invention, that of boring cylinders. This proved critical for the feasibility of the Watt engine. He pushed the wider use of iron for purposes appropriate and inappropriate, from water pipes and barges to pulpits and coffins. His career reflects several of the factors we have mentioned – schooling at a prominent dissenting academy at Kendal – a family background in the industry – apprenticeship to a large Liverpool ironmonger. But he was highly idiosyncratic; Nonconformity did not prevent his cheating Boulton and Watt over the patent rights of their engine or fathering illegitimate children. Like other major industrialists he issued a token coinage, but his was of silver as well as copper, the silver bearing his own head. He did not confine himself to iron, but was an important lead producer, and as an ally of the monopolist Thomas Williams, a considerable force in the politics of the copper trade. From the only permanent element among his father's ventures, the Bersham furnaces in North Wales, he extended operations to the first West Midland coke furnace at Bradley and then gradually assumed control of Willey and New Willey furnaces near Iron-bridge, finally ousting his father from Bersham itself. Late in life he erected another furnace at Brymbo in North Wales. The size of his operations, his advances in blowing, casting and boring techniques and his knowledge of the industry meant that he was much consulted when ironmasters dealt with the government on trade and tariff issues, but he preferred getting on with his business. The Wilkinsons were confident of their superiority and had no worries about foreign experts seeing their plant. Indeed the younger brother, William Wilkinson, spent long periods in France, impart-ing first the family casting and cannon-boring techniques at Indret and then coke smelting itself at Le Creusot. John Wilkinson's drive to expand his concerns by combining technical advance with commercial ambition, his touch of ruthlessness, even megalomania, and his sense of publicity made him celebrated in his own time and subsequently. If not typical, he exhibited in a heightened degree the traits of dynamism and toughness which were coming to be associated with the leading entrepreneurs in iron [14; 15; 16, *Ch. II*].

John Wilkinson's pattern of rather scattered works did not remain typical as the iron industry moved to the use of coal and to locational concentration. It needed to be near to iron ore, but nearness to coal fuel was critical, not only for smelting furnaces, but also for puddling and balling furnaces, for *cupola* furnaces to remelt pig for casting, and for steam engines. Because of the varied nature of ironworks and the difficulty of interpreting old systems of accounting it is not easy to show the way capitals at the disposal of the entrepreneurs grew, but to give one celebrated example, the Coalbrookdale Co., which had a capital of under £3000 in 1708, had a net total value of £165,000 in 1810 [9, *66 seq*].

Some of the largest concerns and capitals were concentrated in South Wales by the early nineteenth century, but we have to bear in mind when talking about ironworks that some might comprise iron and coal mines and extensive systems of industrial railway, so that the whole enterprise extended beyond ironworks themselves. The way in which the geography and resources of the region lent themselves to the creation of large units, together with the lack of local enterprise and capital, meant that capital infusions from merchants coming from outside South Wales were important in its first rapid expanision; in 1806 this accounted for about half the entrepreneurs and half the furnaces [19, *105*]. After these initial infusions the 'ploughing back' of profits into the firm often became the major source for further expansion of capital. But some South Wales ironmasters were particularly dependent on bank loans by 1815, and when the Bank of England opened provincial branches in the 1820s it quickly became involved with the South Wales iron trade [9, *209*; 70, *46*]. The Dowlais works in its initially unsuccessful 1759 partnerhsip had only £400 capital. John Guest from Shropshire joined the firm as manager in 1767, and became a partner in 1782, when the capital was still under £7000. There were three furnaces by 1793, and the capital reached £38,000 in 1787 and £61,000 in 1798. When Josiah John Guest entered the firm in 1806 it made about 5500 tons of iron a year, by 1819 over 10,000 tons; by 1847 its 18 furnaces made 70,000 tons a year and the entire enterprise employed 7000 workers. A few years later its capital valuation passed half a million, to which should be added a further 'floating capital' of £108,000.

The first generation of major ironmasters in South Wales did well because of the low rents which had been asked at the outset for land which had been barren and nearly worthless before they came on the scene, but their successors' profits were less easily made; once early leases fell in greatly increased rents were charged. Cyfartha was an even larger enterprise than Dowlais. The first William Crawshay (1764–1834) put over a quarter of a million pounds into the works, partly to buy out the interest in the concern of those outside his own family, and at his death in 1834 it was already valued at £500,000. The Crawshays remained London merchants in the iron trade even while they were the country's largest ironmasters, and they maintained great investments in the City and in government stock. William II (1788–1867) carried out discounting operations, lent to discount houses and merchant bankers, and had at one time £400,000 in consols. Their financial and mercantile bias seems to have eventually crippled their technological and industrial dynamism [1]. In other regions, capitals in the iron industry were much smaller than those of the large South Wales works. In South Staffordshire there were many small estates with iron and coal mines. Their owners often set up ironworks of lesser size on the basis of their minerals, aided by investments from tradesmen in neighbouring towns and by managers bought in from the staff of existing ironworks. They were helped by the accessibility of their local resources; it was not unknown for coal, iron ore and fireclay to be wound up the same shaft.

Clearly in the first half of the nineteenth century the iron industry was marked by large capitals, massive investment in plant and large workforces as compared with most other industries. There is room, however, for much more research on the ironmasters of the eighteenth and early nineteenth centuries, to elucidate and compare their roles as business and as technological innovators; it is only those who were successful in both capacities who tend to have attracted the attention of historians. It may be worth observing that the great technical breakthroughs generally came from men who were not already established ironmasters. Darby, who introduced coke smelting, had earlier been involved in the making up of iron goods and the production of non-ferrous metals; Huntsman, who established quality steelmaking methods which held the field for over a century, was a clockmaker; Cort, who gave the industry the method of making wrought iron from pig which lasted as long as demand existed, was a commercial man who entered ironmaking accidentally; Crawshay, who saw through the critical 'R & D' phase of that invention was a merchant; Neilson was a gas engineer.

(i) IRONMASTERS' ASSOCIATIONS

As in some other important industries, like coal and copper, major ironmasters, even some of those most celebrated as leaders in business expansion and innovation, were strong proponents of a regulated rather than a free trade in their product.

The West Midlands ironmasters were certainly meeting periodically to fix prices in the 1730s and 1740s [71]. The Darbys were in an arrangement with the Wilkinsons to set the prices for an important range of castings in the early 1760s; Midlands ironmasters were meeting on a regular quarterly basis at Stourbridge by the mid 1770s. In the 1780s the Midland ironmasters were important in the

Birmingham Commercial Committee, and in the next few years were involved in petitioning and making representations, with their colleagues in other parts of the country, against a proposed excise on pig iron, and against a proposed duty on coal. From the 1770s they opposed attempts to harmonise import duties on iron entering England with the lower ones ruling for Ireland, and then for similar reasons they joined in the opposition to the commercial proposals of 1785 for liberalising trade between England and Ireland.

The ironmasters' presence was significant in the General Chamber of Manufacturers which had a brief importance at this period, and before it became divided and ineffective they had been confident enough of their own competitive position to be in that section of the Chamber welcoming the lower tariff duties between England and France achieved in the short-lived Eden Treaty of 1786. Their liberalism in matters of trade was an expression of self-interest rather than of any free trade ideal; most were firmly in favour of the laws against the exportation of tools and machinery, and they were not keen to see the new technology of the early industrial revolution readily made available to foreign competitors. Once again in 1796 the ironmasters got together to make sure that wartime financial pressures did not result in the government placing duties on coal and iron, and in the same way they fought off an iron duty in 1802. By this time the Midland ironmasters met frequently in Birmingham, and the South Wales ones in Newport, and at different times in the early decades of the nineteenth century meetings of groups of ironmasters are known of for Staffordshire, Shropshire, North Wales and Scotland, though some groups were not constantly in existence, or were eventually absorbed in others. The papers of an association of Derbyshire and Yorkshire ironmasters survive from the years 1799–1828. Pig prices were being concerted between them from 1800, and there were agreements on the prices of castings, and in 1809 for bar. From 1808 they met with the Welsh, Shropshire and Staffordshire ironmasters fairly regularly at Gloucester, so that there was for a time an attempt at something like a national price agreement.

The post-war depression and the subsequent years of price fluctuation made it impossible to make effective national agreements, even some of the regional associations became impermanent or weak, and agreed curtailment of production was tried, at least in South Wales. Secret discounts or the giving of unusually long credits by weaker or greedier firms were a problem in normal times, a very serious one in depressions [3]. National agreement was again sought, but not effectively, at Gloucester meetings in 1825 and 1831. While the verdict on national agreements between the late Napoleonic Wars and the early 1830s seems to be that the intense competition that ironmasters frequently found themselves in meant that such agreements had no great effect, they testify to a strong inclination towards control of prices and even of output. In the late 1830s a downturn of trade led to a more serious attempt to control output which may have been implemented in 1836, but was certainly achieved by a 20 per cent reduction in 1840. An 1842 agreement brought in Yorkshire, Derbyshire, Staffordshire and South Wales, though the last was weak in allegiance; Scotland, at first pessimistic and not convinced of its possibility, also eventually joined. Even so some large concerns, including those of Crawshay in Wales and Dixon in Scotland, did not come in [3, *Ch. VIII*; 9, *Ch. VI*]. However, this relative unity, though temporarily strengthened by the severe recession of 1839–42, did not last, especially once good times returned bringing with them the unexampled demand for railway iron. The iron trade increasingly conformed to the prevailing competitive and free trade atmosphere of the mid-century years, only again turning to plans to effect a national association, as did the leading firms in some other industries, in the 1870s. Though before the mid 1840s there had been incentives to concert trade policy regionally and nationally there were always forces working against such cooperation. There was some regional specialisation in types and quality of iron, so markets were not identical and could be differently affected by general movements of demand; ironmasters who owned their own coal or iron ore mines were differently placed from those who did not, and could sacrifice the profits of some of their undertakings to compensate those of others. Those with large capitals and important in marketing, like the Crawshays, could hold stocks when others were unable to, even if this might be a dubious strategy in the longer term. Finally there were always the smaller producers, or those whom technical or commercial inefficiency had left living from hand to mouth, who were difficult to bring into any agreement, and were readily tempted to break any they temporarily entered into [9; 1; 43, *126–7*].

In principle this brief account of the ironmasters should be balanced by one of the ironworkers. However, this pamphlet has confined itself to the primary iron industry and some accounts of labour in the iron industry also consider those making up goods in iron and steel, and workers in iron company coalmines. Though there is a good deal of material for the last decades of our period and centred on a few very large concerns, it would not be possible to give a coverage of any chronological evenness. A modern study of the history of the ironworker with a long historical perspective is overdue. Ashton's account, based on his work of 60 years ago, has much on workers other than those in primary iron production, but Birch provides some good material for the last decades of our period [3, *Ch. IX*; 9, *244–73*] and Addis [1] is useful for South Wales.

(ii) DECISIVE DEPARTURES

The end of our period, 1850, is an appropriate time to close this survey. In the next decade was to begin the first erratic introduction of the Bessemer process, heralding the age of cheap steel, with its supersession of wrought iron for uses like rails, for the plate used in building ships (which had begun to be a significant market for wrought iron as our period ends) and for constructional purposes. The Bessemer process was originally tied to haematite iron and this led to important adjustments in ore markets, but Bessemer's was not the only successful process, it was swiftly followed by the open-hearth method, and both were suitable for a wide range of ores once the ability to use phosphoric ores was achieved by the Gilchrist-Thomas invention. In the same decades came the introduction of special steels with important implications for engineering uses. While as late as 1873 Britain produced as much pig iron as Europe and the USA together, the iron and steel industry was now carried out at a high technical level in the leading European countries and in the United States, and British primacy, both in output and in innovation, was, after a long period of leadership, about to decline.

APPENDIX 1 Industrial Archaeology

The history of industry can be approached by the student not only through published texts, but also through industrial archaeology, the study of the industrial past through its physical remains. These can sometimes be found in conventional museums, but in recent years outstanding historical industrial sites and their equipment have been preserved and exhibited, and these have sometimes then served as a focus for the gathering-in of important industrial remains originating on other sites, which it has not been possible to preserve in their original situation. Visits to such museums are excellent aids to teaching and learning, fixing the scale, nature and methods of working of industrial equipment firmly in the mind, and impressing the historical geography and architecture of classic industrial sites on the observer. The iron and steel industries are comparatively well served in this respect.

The best known conservation project connected with the iron industry is the Ironbridge Gorge Museum in Telford, Shropshire, where the old furnace where Abraham Darby first smelted iron with coke in 1709 is conserved together with the Bedlam furnaces of the late 1750s and the Blists Hill furnaces of the 1830s with their attendant engine houses. The museum has recreated a Wrought-Iron Works within the Blists Hill Open Air Museum where puddling furnaces, rolling mills and a steam hammer can be seen at work. Blists Hill also includes a working foundry, and artefacts relating to the iron industry are displayed in the Museum of Iron at Coalbrookdale. The outstanding charcoal-using blast furnaces which survive are at Charlcott in south Shropshire, Bonawe in Argyll and Duddon in the Lake District. The best concentration of early coke-using blast furnaces is in South Wales where the sites at Blaenavon, Clydach, Ynysfach in the middle of Merthyr, Hirwaun and Neath Abbey are all worth visiting. There are substantial remains at Dowlais of the blowing house and stable block of one of the largest of nineteenth-century iron works. Two short-lived blast furnaces of the early nineteenth century which are well preserved are those at Whitecliffe in the Forest of Dean and Moira on the Derbyshire/Leicestershire border. In the Telford area there are substantial remains of furnaces of the 1820s at Stirchley and Hinks Hay. Nothing substantial remains of any traditional finery and chafery forge, although there are some fragments of the forge at Stoneyhazle in the Lake District and many sites in Staffordshire, Worcestershire, Shropshire and the Forest of Dean have remains of water-power installations. Wortley Top Forge near Sheffield retains water-powered hammers of the type used in finery and chafery forges, although it was converted in the nineteenth century into a plant for the manufacture of railway axles. The Abbeydale Industrial Hamlet run by the Sheffield Museum Service on the southern edge of the city includes a small crucible steel works as well as water-powered hammers. A cementation furnace and the remains of another are preserved in the centre of Sheffield but the best furnace of this kind is that at Derwentcote, south of Newcastle-upon-Tyne, which is currently being restored by English Heritage.

Industrial archaeology generally as well as the iron and steel industry in particular is well treated in Neil Cossons' *The BP Book of Industrial Archaeology* (second edition, David and Charles, 1987) or R. A. Buchanan's *Industrial Archaeology in Britain* (Penguin, 1972). The historical interrelation between industry and its environmental background is best described in Barrie Trinder, *The Making of the Industrial Landscape* (Dent, 1982, paperback edition Alan Sutton, 1987) which is particularly good on this aspect of the iron industry.

APPENDIX 2   Iron Production

The Output of Pig-iron in Great Britain by Decades, 1530–9 to 1860–9

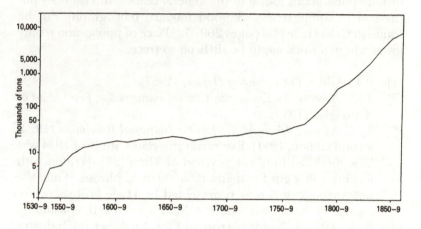

SOURCE: [69].

Bibliography

While indicating the works cited in the text, and endeavouring to include publications useful to the general reader, this list does not pretend to completeness. A good modern bibliography can be found in C. K. Hyde [43] pages 260–72. Place of publication is only given where a work might be difficult to trace.

[1]   J. P. Addis, *The Crawshay Dynasty* (1957).

[2]   T. S. Ashton, *An Eighteenth Century Industrialist: Peter Stubs of Warrington* (1939).

[3]   T. S. Ashton, *Iron and Steel in the Industrial Revolution* (1924, second edition, 1951). Essentially this classic study of 1924 was little modified until the revision of Flinn [23]. Written with lucidity and a gift for memorable turns of phrase. While the new orthodoxy is now represented by Hyde [43] there are aspects of the subject where Ashton is still worth consulting.

[4]   E. E. Aström, 'Swedish Iron and the English Iron Industry about 1700: Some Neglected Aspects', *Scandinavian Economic History Review*, XXX (1982).

[5]   B. G. Awty, 'Charcoal Ironmasters of Cheshire and Lancashire, 1600–1785', *Transactions of the Lancashire and Cheshire Historic Society*, CIX (1957). A model regional study.

[6]   B. G. Awty and C. B. Phillips, 'The Cumbrian Bloomery Forge in the Seventeenth Century and Forge Equipment in the Charcoal Iron Industry', *Trans. Newcomen Soc*, 51 (1979–80).

[7]   K. C. Barraclough, *Steelmaking Before Bessemer: Vol. I, Blister Steel, the birth of an industry; Vol. II, Crucible Steel, the growth of technology* (1984). Quite outstanding, particularly on the technical side. Good material on comparative European techniques; excellent documentary excerpts especially from European sources.

[8]   J. F. Belhoste, 'L'Inventaire des forges françaises et ses applications', *Ironworks and Iron Monuments*, Ironbridge Symposium (ICCROM, Rome, 1985).

[9]   A. Birch, *The Economic History of the British Iron and Steel Industry 1784–1879* (1967). A little dated in some respects, but a good scholarly work covering a period of exceptional growth in the industry.

[10]   N. Blundell, *The Great Diurnall of Nicholas Blundell of Little Crosby, Lancashire*, vol. 2 1712–19 (ed. F. Tyrer and J. J. Bagley, *Lancs. and Chesh. Record Soc.*, 1970).

[11]   J. Butt, 'The Scottish Iron and Steel Industry before the Hot Blast', *Journal of the West of Scotland Iron and Steel Institute*, LXXIII (1965–6). A very useful and clear account.

[12]   R. H. Campbell, 'The Financing of Carron Company', *Business History*, 1 (1958).

[13]   R. H. Campbell, *Carron Company* (1961). A valuable business history of a great works which introduced the new coke-smelting technology to Scotland.

[14]   W. H. Chaloner, 'Les frères John and William Wilkinson et leurs rapports avec la métallurgie française (1775–1786)', *Annales De L'Est*, mémoire 16 (Nancy 1956). An account of the transfer of the new British technology to France which deserves to be better known.

[15]   W. H. Chaloner, 'Isaac Wilkinson, Potfounder' in L. S. Pressnell (ed.) *Studies in the Industrial Revolution* (1960).

[16]   W. H. Chaloner, *People and Industries* (1963). Has a short but useful chapter on John Wilkinson.

[17]   H. Cleere and D. Crossley, *The Iron Industry of the Weald* (1985). A very thorough study which sets out the results of many years archaeological and documentary research on this area.

[18]   D. W. Crossley, 'The Management of a Sixteenth Century Ironworks', *Economic History Review*, Second Series, XIX (1966).

[19]   F. Crouzet, *The First Industrialists* (1985). Has some very good material on the ironmasters of the early industrial revolution.

[20]   Joan Day, 'The Continental Origins of Bristol Brass', *Industrial Archaeology Review*, VII, No. 1 (1984).

[21]   A. Den Ouden, 'The Production of Wrought Iron in Hearths', *Journal of the Historical Metallurgy Society* (1981).

[22]   R. L. Downes, 'The Stour Partnership, 1726–36', *Economic History Review*, Second Series, III (1950).

[23]   M. W. Flinn, 'The Growth of the English Iron Industry, 1660–1760', *Economic History Review*, Second Series, XI

(1958). A very important article which constituted the first major revision of Ashton's classic work.

[24] M. W. Flinn, 'William Wood and the Coke-Smelting Process', *Trans. Newcomen Soc.*, XXXIV (1961–2).

[25] M. W. Flinn, *History of the British Coal Industry* 2 (1984).

[26] W. K. V. Gale, 'Wrought Iron: A Valediction', *Trans. Newcomen Soc*, XXVI (1963–4). Excellent on the craft aspect of the process originating with Cort.

[27] W. K. V. Gale, *The British Iron and Steel Industry, A Technical History* (1967). Gale has a remarkable ability to make technology understandable to the non-technical reader.

[28] W. K. V. Gale, *The Iron and Steel Industry: A Dictionary of Terms* (1971). Contains not only modern terms but many now obsolete relating to the industry in earlier periods.

[29] W. K. V. Gale, *The Black Country Iron Industry: A Technical History* (1979).

[30] R. B. Gordon and T. S. Reynolds, 'Mediaeval Iron in Society', *Technology and Culture*, 27 (Jan. 1986).

[31] T. R. Gourvish, *Railways and the British Economy 1830–1914* (1980). In the same series as this pamphlet. It has a section on the economic relationship between railways and the iron industry which is more specialised than the account given here.

[32] G. Hammersley, 'The Charcoal Industry and its Fuel, 1540–1750', *Economic History Review*, Second Series, XXVI (1973). An important paper which attempts a major revision of earlier assumptions.

[33] J. R. Harris, 'The Introduction of coal into iron smelting', *Edgar Allen News*, 37 (1958), 435.

[34] J. R. Harris, 'Industry and Technology in the Eighteenth Century: England and France', printed lecture (Birmingham, 1972).

[35] J. R. Harris, 'Skills, Coal and British Industry in the Eighteenth Century', *History* (1976). Takes some of its examples from the iron and steel industries.

[36] J. R. Harris, 'Attempts to Transfer English Steel Techniques to France', in S. Marriner (ed.), *Business and Businessmen* (Liverpool, 1978).

[37] C. Hart, *The Industrial History of Dean* (1971).

[38] G. R. Hawke, *Railways and Economic Growth in England and*

Wales (1970). An important and strongly argued work using the techniques of the modern economic historian.

[39] K-G. Hildebrand, 'Foreign Markets for Swedish Iron in the Eighteenth Century', *Scandinavian Economic History Review*, VI (1958). A most useful and constructive paper of great relevance to the study of the British industry, which has not been superseded.

[40] C. K. Hyde, 'The Adoption of Coke Smelting by the British Iron Industry, 1709–1790', *Explorations in Economic History*, X (1972–3).

[41] C. K. Hyde, 'The Adoption of the Hot Blast by the British Iron Industry, A Reinterpretation', *Explorations in Economic History*, X (1972–3).

[42] C. K. Hyde, 'Technological Change in the British Wrought Iron Industry, 1750–1815: A Reinterpretation', *Economic History Review*, Second Series, XXVII (1974).

[43] C. K. Hyde, *Technological Change and the British Iron Industry 1700–1870* (Princeton, 1977). A key work, as much the historical orthodoxy of the 1980s as Ashton's book was that of the middle decades of the century. It embodies the revisionary studies of the previous three articles.

[44] L. Ince, *The Neath Abbey Iron Company* (Eindhoven, 1984).

[45] A. H. John and Glanmor Williams, *Glamorgan County History Vol. V; Industrial Glamorgan from 1700 to 1970* (1980).

[46] B. L. C. Johnson, 'The Foley Partnership; The Iron Industry at the end of the Charcoal Era', *Economic History Review*, Second Series, IV (1952). A classic article.

[47] B. L. C. Johnson, 'The Midland Iron Industry in the Early Eighteenth Century', *Business History*, II, No. 2 (1960). Has some natural overlap with the article above; a most useful regional study.

[48] J. Kanefsky and J. Robey, 'Steam Engines in Eighteenth Century Britain; A Quantitative Assessment', *Technology and Culture*, 21 (1980).

[49] G. Magnusson, 'The Mediaeval Blast Furnace at Lapphyttan' in *Iron Works and Iron Monuments*, Ironbridge Symposium (ICCROM, Rome 1985).

[50] B. R. Mitchell, 'The Coming of the Railway and United Kingdom Economic Growth', in M. C. Reed (ed.), *Railways in the Victorian Economy* (1969).

[51] B. R. Mitchell and P. Deane, *Abstract of British Historical Statistics* (1962, reprinted 1971 and 1976).

[52] G. R. Morton and N. Mutton, 'The Transition to Cort's Puddling Process', *Journal of the Iron and Steel Institute*, CCV (1967). Re-established the importance of the pre-Cort processes.

[53] G. R. Morton and M. D. G. Wanklyn, 'Dud Dudley – A New Appraisal', *Journal of West Midlands Regional Studies*, 1 (1967).

[54] R. A. Mott, 'Dud Dudley and the Early Cast Iron Industry', *Trans. Newcomen Soc.*, XV (1934–5).

[55] R. A. Mott, 'The Earliest Use of Coke for Iron Making', *Gas World* (1957).

[56] R. A. Mott, 'Abraham Darby (I and II) and the Coal-Iron Industry', *Trans. Newcomen Soc.*, XXXI (1957–9).

[57] R. A. Mott, 'The Coalbrookdale Horsehay Works: Part I', *Trans. Newcomen Soc.*, XXXI (1957–9) and 'Part II', XXXII (1959–60).

[58] R. A. Mott (ed. Peter Singer) *Henry Cort: The Great Finer, Creator of Puddled Iron* (1983). Mott's writings are not easy reading and he was very tenacious of his opinions, but his careful research led to many revisions. The Cort book is the most readable and gives the latest account of a most important inventor.

[59] L. B. Namier, 'Anthony Bacon, M.P., an eighteenth century merchant', *Journal of Economic and Business History*, II (1929).

[60] J. Needham, 'Iron Production in Ancient and Mediaeval China', *Trans. Newcomen Soc.*, XXX (1956–7).

[61] J. Needham, *The Grand Titration; Science and Society in East and West* (1969).

[62] J. U. Nef, *The Rise of the British Coal Industry*, 2 vols (1932, reprinted in 1966). A great work which established new standards for research in its field. Some of its strongly advocated ideas – like that of an earlier industrial revolution – have fallen from favour, but its emphasis on the linkage between coal and major long-term changes in technology and economy is its essential feature and remains of perennial interest.

[63] P. O'Brien, *The New Economic History of the Railways* (1977).

[64] C. B. Phillips, 'The Cumbrian iron industry in the seventeenth century' in W. H. Chaloner and Barrie M. Ratcliffe (eds), *Trade and Transport*, Essays in Economic History in

honour of T. S. Willan (1977).

[65] A. Raistrick, *Quakers in Science and Industry* (1950 and 1968).

[66] A. Raistrick, *Dynasty of Ironfounders; The Darbys and Coalbrook-dale* (1953). This celebrated book on the ironmasters in whose works a series of great technological advances took place, and where Quaker influence was pervasive, has been written by a Quaker historian who has made great contributions to the history of industry and technology and to that of the Society of Friends.

[67] A. Raistrick and E. Allen, 'The South Yorkshire Ironmasters 1690–1750', *Economic History Review*, IX (1939).

[68] J. E. Rehder, 'The Change from Charcoal to Coke in Iron Smelting', *Journal of the Historical Metallurgy Society*, 21, No. 1 (1987). A new view on the introduction of coke smelting, founded on technical experience at American furnaces. The capacity of this historical technical breakthrough to attract new scholarship and further reinterpretations seems inexhaustible.

[69] P. Riden, 'Output of the British Iron Industry before 1870', *Economic History Review*, Second Series, XXX (1977), A good survey article with useful statistics. Makes much use of the earlier work of Hyde and Hammersley.

[70] R. O. Roberts, 'The Operations of the Brecon Old Bank of Wilkins & Co. 1778–1890', *Business History*, I (1958).

[71] M. B. Rowlands, *Masters and Men in the West Midland Metalware trades before the Industrial Revolution* (1975). Good on the organisation of trade and the interrelations of all those involved.

[72] H. R. Schubert, *History of the British Iron and Steel Industry, from c. 450 B.C. to A.D. 1775* (1957). Much more technological than economic in its approach, it is a very scholarly work which still remains a valuable source.

[73] E. B. Schumpeter, *English Overseas Trade Statistics* (1960).

[74] R. G. Schafer, 'Genesis and Structure of the Foley "Ironworks in Partnership" of 1692', *Business History*, XIII, No. 1 (1971).

[75] A. Slaven, *The Development of the West of Scotland 1750–1960* (1975). Sets much of the Scottish iron industry very carefully into its economic background.

[76] S. Smiles, *Industrial Biography: Iron workers and Tool makers* (1863).

[77] S. Smith, *A View from the Iron Bridge* (1979). This is a well-

illustrated list of mainly eighteenth- and early nineteenth-century drawings, landscapes and portraits depicting the industrial activities, and some of the industrial leaders, of the Severn Gorge. It gives many illustrations of the great bridge itself, and shows how the bridge and its setting captured the imagination of contemporaries.

[78] E. Straker, *Wealden Iron* (1931, reprinted 1969). A pioneering work making an interesting contrast with the new study of Cleere and Crossley.

[79] Thomas, Brinley, 'Was there an Energy Crisis in Great Britain in the 17th Century?', *Explorations in Economic History*, 23 (1986). Reopens the debate on fuel availability and price largely rehabilitating the Nefian case [62].

[80] J. M. Treadwell, 'William Wood and the Company of Ironmasters of Great Britain', *Business History*, XVI, No. 2 (1974).

[81] B. Trinder, *The Industrial Revolution in Shropshire* (1973, second edition 1981). A fine regional study putting an up-to-date account of the local iron industry into the framework of other industrial activities, transport development and social life.

[82] B. Trinder, *The Making of the Industrial Landscape* (1982).

[83] R. F. Tylecote, *A History of Metallurgy* (1976). Valuable for those wishing to look in more detail at the metallurgy aspect of the history of the metal industries.

[84] R. F. Tylecote, *The Prehistory of Metallurgy in the British Isles* (1986). Contains some important revisions of the previous work.

Index